PEST CONTROL IN BUILDINGS

A guide to the meaning of terms

THE RENTOKIL LIBRARY

Pest control in buildings covers many facets: building construction and transport; control methods with chemicals applied in different ways (e.g. as a fog); a knowledge of the habits of pests and careful inspection; protection of the public against diseases spread by pests, especially where food may be contaminated.

In these photographs, and others throughout the book, the author wishes to make it clear that no criticism of the premises or foodstuffs is intended; they are included only as illustrations of pest control problems.

PEST CONTROL IN BUILDINGS

A guide to the meaning of terms

P. B. CORNWELL

Director of Research
Rentokil Limited

HUTCHINSON OF LONDON

HUTCHINSON & CO (*Publishers*) LTD
3 Fitzroy Square, London W1

London Melbourne Sydney Auckland
Wellington Johannesburg Cape Town
and agencies throughout the world

First published 1973

69·0591

Produced by Hutchinson Benham Ltd

*This book has been set in Times type, printed in Great Britain
on art paper by Anchor Press, and
bound by Wm. Brendon, both of Tiptree, Essex*

ISBN 0 09 114130 3

CONTENTS

PREFACE

At the Conference of the British Pest Control Association in Jersey in October 1971, representatives of industry and government from eighteen nations met to discuss pest control in the decade ahead. It was at this meeting that my colleague, Dr. Norman Hickin, suggested I write this book. He chose the right moment. When many people were attempting to communicate in common terms, it seemed of value to put those terms on paper—in the text of this book—to extend international understanding of the subject.

ACKNOWLEDGEMENTS

Two people who have been part of commercial pest control for many years have helped me in the production of this book. Considerable assistance has been given by my colleague, Robin Edwards, with suggestions on how best to present the information in the text. He has corrected and edited drafts for technical content, taken many of the photographs which appear as colour plates, arranged them in suitable groups and designed the book jacket. I am deeply indebted to him for the constant support which has lightened the task of getting this volume published.

I am equally grateful to Miss Shirley Edwards for her day-to-day secretarial help in the production of this book. She has typed all the drafts and the final text with care and enthusiasm.

Others at the Laboratory of Rentokil to whom I am indebted for assistance are my colleagues John Bull, David Lynch, David Moore and Brian Ashworth, each of them specialists in one or other branch of pest control.

Many friends in a number of organisations have also contributed. I would particularly like to thank Mr. Dennis Papworth (M.A.F.F.), Mr. F. H. Jacob (P.I.C.L.) and Dr. Ralph Heal (N.P.C.A.) for their help.

ABOUT THIS BOOK

Pest control in buildings has many facets. The subject concerns itself with the pests themselves, their biology, aspects of public health, pest control chemicals, methods of applying them and equipment. It involves food, manufacturing processes, aspects of building construction, toxicity and legislation. It involves people: those who have pest problems, technical staff in research departments of industry and government, those who apply pesticides and public opinion.

Pest control is a complex subject which has made great strides in recent years from the activities of the novice to the need for professionalism, with a technical and scientific approach. Two demands have accelerated this change: 1) that of a modern society for clean wholesome food and pest-free surroundings; and 2) that pest control shall in no way put man and his environment at risk through the methods, mainly chemical, that are used.

The student of pest control is a phenomenon of the last 15 years. The existence today of training programmes to equip men for a career in pest control is an indication that the subject requires a combination of technical competence and practical experience; an ability to understand and use technical terms and to make use of technical information.

People with pest problems require the help of men with a thorough knowledge and understanding of their subject—also with the practical ability to do the job. Equally, government bodies require the pest control industry to present its case in an informed scientific manner supported adequately with factual information. This book is intended as a contribution to these objectives.

This guide to the meaning of terms provides a reference to the 'words of the industry' of both past and present. The major difficulty has been to decide what to leave out. Many chemicals have largely disappeared from use in recent years through more effective and safer alternatives becoming available and removal or reduction in use through legislative control. However, it was thought desirable to include reference to many of the older and well-tried pesticides, since the degree of sophistication of pest control varies considerably in different parts of the world and so too does the scope of government regulations.

This is not a dictionary definition of terms; rather the book gives a thumb-nail sketch of the implications behind the 'word', especially its interrelation with other aspects of pest control. It is hoped that the extensive cross-referencing in the text will be found an advantage.

One of the purposes has been to produce a companion volume to Norman Hickin's *Wood Preservation—A guide to the meaning of terms*; the two books thus providing a source of reference to the subject of pest control in buildings, including also the use of preserved wood. The presentation in the two volumes has been kept as similar as possible in the alphabetical referencing, and with the listed words categorised under one or more abbreviations as shown below. In some instances the books overlap; this is desirable and inevitable, since some terms have been re-defined, but in a different context.

The disciplines used in classifying terms are as follows:

bldg	Building.
chem	Chemical.
dis	Disease and causative organism.
ent	Entomological.
equip	Equipment used in pest control.
leg	Legislation and other approval schemes.
manuf	Manufacturing.
name	Name of company, organisation or association, or a person in pest control or special activity.
phy	Physical.
proc	Process or method.
p. prod	Proprietary product.
ship	Shipping.
tox	Toxicological.
zoo	Zoological.

In deciding the scope of this book, the control of four pests has been specifically omitted. Three present entirely outdoor activities, namely the control of weeds, moles and pests of turf. The fourth, the control of fungi which attack building timbers, is adequately covered in the companion volume.

All industries tend to develop their own jargon involving words or phrases quite unintelligible to others. Examples are 'OPs', 'clean out', 'call-back'; a building 'alive' with cockroaches, doing a 'fogging job'; 'spot treatment'. These examples indicate lack of precision in thinking and laziness in our efforts to communicate.

A common reason for poor communication is the difference in inter-pretation which people, inside and outside the pest control industry, place on words. The simplest example is the term 'infestation'. It may suggest to some that an infested building contains a very large population or is overwhelmed with cockroaches, but to the manufacturer who has a few cockroaches in his building, 'infestation' may appear to be the wrong term to describe the conditions.

An example of this problem arose in connection with the Prevention of Damage by Pests Act, 1949. During the passage of this Bill through the House of Lords, one of the noble lords rose to his feet and said: 'Hickory Dickory Dock, the mouse ran up the clock—is that an infestation?' An effort, quite rightly, to seek an agreed definition of the word to be used in U.K. law. It is to avoid doubt in the use of pest control terms that this book has been written.

A

Abate. (*p. prod*) An organophosphorus compound, being evaluated for the control of mosquito larvae at extremely low rates of application, offering promise of use without harm to fish.

The acute oral LD_{50} (rat) is 2300 mg/kg; relatively non-toxic to bees, birds and aquatic organisms. Formulations available include emulsion concentrates, wettable powders and granules.

Abdomen. (*zoo*) The hind part of animals, carrying the anal and genital openings and containing the terminal sections of the gut, the reproductive system and most of the excretory organs. In insects, the segments behind the thorax: there are basically 11 segments, but this number is often reduced, and some are telescoped so that they are invisible externally. Most visible segments carry spiracles on each side opening into the respiratory system.

In adult insects, appendages are usually absent, except sometimes for a pair of anal cerci and the genital appendages of which the most obvious are the claspers (male) and ovipositor (female).

In woodlice, millipedes and centipedes, most abdominal segments carry legs or leg-like appendages. In mammals, the abdomen is not defined externally, but consists of that part of the body lying between the diaphragm and the pelvis.

Absorption. (*phy*) Of insecticides, the passage of all or part of a spray into the surface to which it is applied. Of fumigants, the passage of the gas into the commodity being fumigated. Dermal absorption; the entry of a pesticide into the body via the skin following accidental contamination. Into insects, the passage of insecticide through the cuticle.

Acanthomyops. (*ent*) See Lasius.

Acanthoscelides obtectus. (*ent*) Coleoptera: Bruchidae. The Bean beetle, Bean weevil. Breeds on all types of stored pulses. Similar in habits to Callosobruchus maculatus (*q.v.*). The adult (3 mm long) is olive-brown with darker brown marks on the elytra; thorax covered with fine yellow-orange hairs. Eggs are laid singly on or near beans but not glued to the surface. Reproduction ceases below 15°C.

Acari. (*zoo*) Acarina. The subclass or order of Arachnida containing mites and ticks.

Mites have some similarities with insects in that they have jointed legs (6 in larvae and 8 in adults) and an exoskeleton, but their bodies are without division into head, thorax and abdomen. Mites appear to the naked eye as coarse dust and can often only be detected by their movement. They infest a variety of materials; e.g. cereals, flour, cheese and smoked meats. The species of economic importance are found principally in mills, grain and cheese stores, and domestic larders where they attack any foodstuff with a sufficiently high moisture content. Mites (particularly *Acarus siro*) taint food and produce a characteristic 'minty' odour. They can also cause dermatitis to those handling infested food, some species causing severe skin irritation (e.g. the Furniture mite, *Glycyphagus domesticus*). Other troublesome mites are the House dust mite, *Dermatophagoides pteronyssinus*, the Itch mite, *Sarcoptes scabiei*, and the Harvest mite, *Trombicula autumnalis*, which cause allergic reactions in lungs and on the skin. See also *Dermanyssus gallinae* and *Tyrophagus*.

Mites are readily killed by high temperatures and low humidities. Where possible the drying of foodstuffs is the most effective method of control.

Ticks (order Metastigmata or family Ixodidae) are much larger than mites and all feed on vertebrate blood. They are occasionally brought into the house on domestic animals (e.g. the Sheep and cattle tick, *Ixodes ricinus*).

Acaricide. (*chem*) A substance which kills Acari (mites).

Acarus siro. (*zoo*) Acari: Tyroglyphidae. Previously known as *Tyroglyphus farinae*, the Flour mite is a serious pest of cereals and cereal products when stored at high moisture contents. Females lay a few eggs per day with a total of 30 to 50; they are similar to the eggs of other mite species and hatch in 3–4 days. The larva is minute (0·15 mm) with three pairs of legs; it moults after 4–5 days into the 1st stage nymph with four pairs of legs. The 2nd stage nymph moults to adult; there is no pupal stage. A hypopus (resting stage) may occur between the nymphal stages. The adult is 0·5 mm long, white or pale brown, with four pairs of legs and is slow moving. The male has a distinct tooth under the front coxae which is a useful identification feature. The minimum period from egg to adult is 9–11 days at 23°C and 90% relative humidity but is much extended under unfavourable conditions.

Accelerated test. (*proc*) A laboratory procedure by which a chemical reaction is speeded up to obtain information more rapidly than in normal practice, e.g. to assess the life, or rate of degradation at high temperature of insecticides applied to test surfaces. A test in which conditions are so arranged as to simulate in a short time the effects of a more prolonged period of ageing.

Access panel. (*bldg*) A removable section of a ceiling, wall, or floor, having the same purposes as a trap door, to allow examination of voids and water, electric and other services. Important also in providing access for pest control purposes. See also FALSE CEILING and FALSE FLOOR.

Acetylcholinesterase. (*chem*) See CHOLINESTERASE.

Acheta. (*ent*) See CRICKETS.

Acute dose. (*tox*) An amount of a substance taken or administered over a short period of time. Of rodenticides, the amount, for example, of a bait consumed by a rat in one feed, or of a test substance given by oral dosing. The use of this term often implies that the amount consumed is highly toxic, but this is not necessarily so.

Acute poison. (*chem*) A toxic substance which is used at a sufficiently high concentration to bring about rapid symptoms of poisoning and death in a relatively short period of time (cf. CHRONIC POISON).

Examples of commonly used acute rodenticides are sodium monofluoroacetate, fluoracetamide, thallium sulphate, zinc phosphide and alphachloralose. The majority suffer the defect of causing discrimination against baits (see BAIT SHYNESS) and sub-lethal feeding by a proportion of the rodent population (see PRE-BAITING).

Additive. (*chem*) A substance incorporated into a pesticide formulation to improve its performance. An example, (bait additive), is the incorporation of flavouring agents into baits to improve their palatability to rodents.

An example, for insecticides is the addition of dichlorvos in small amounts to residual insecticides to improve knockdown or provide flushing action.

Adhesives. (*chem*) See STICKY BOARD.

Adsorption. (*phy*) The retention of a pesticide as a liquid or vapour by a surface, such as plaster or brick, the subsequent rate of evaporation depending upon the form of adsorption. Cf. ABSORPTION.

Adult. (*zoo*) The mature, or reproductive stage of development of an animal. Of insects, the final stage of development showing the external form and coloration characteristic of the species. Previously known as the imago, pl. imagines.

Aëdes. (*ent*) Diptera: Culicidae. A large genus of mosquitoes, many species occurring in Britain. Some are confined to fresh water, others are pests around salt marshes laying eggs in damp soil, and a third group lay eggs in entrapped rain water in artificial containers and holes in trees. Many inflict painful bites outdoors. Commonly occurring species in Britain include *A. rusticus* and *A. cantans* usually near woodland with one generation each year, and *A. detritus* and *A. caspius* breeding in brackish water with several generations per annum.

The principal *Aëdes* mosquitoes in the U.S. are:

Aëdes vexans: a major pest, widespread in all the northern states, breeding in any temporary pool of fresh water. Eggs are laid on the ground and hatch during flooding. *A. vexans* has been recorded in Britain.

Aëdes trivittatus: generally found in the northern States. Larvae occur in woodland pools, the older stages feeding on vegetation on the bottom.

Aëdes sticticus: more abundant in the northern States than in the south. Eggs are laid on stream and river banks and hatch after flooding in the Spring. The eggs may be dormant for 2–3 years. *A. sticticus* has been recorded in Britain.

Aëdes sollicitans: the salt marsh mosquito, one of the most severe pest mosquitoes known in some coastal areas of America. Eggs are laid in mud around brackish waters and must remain dry for 24 hours before they will hatch on wetting by high tide.

Aëdes aegypti: the yellow fever mosquito, a pest in many tropical countries of the world, carrying the viruses of yellow fever and dengue. It breeds in tree holes and water entrapped in roof gutterings and containers. Eggs are laid above the water line, the larvae normally hatching in 4 days. *A. aegypti* may bite at any time of day.

Aerogel. (*chem*) Silica aerogel, 'Drie-die', 'Drione'. A dust which is virtually non-toxic to man but which kills insects by abrasive or absorptive action on insect cuticle, the insect losing water through the wax layer of cuticle. Not as popular as dusts incorporating synthetic insecticides, but aerogels have been used for the control of cockroaches, fleas, mites and ticks.

Aerosil. (*chem*) A fumed silica, sub-micron in particle size, used as an anticaking and antisettling agent in certain formulated pesticides. It is also used as a thickening agent by the paint industry.

Aerosol. (*phy*) A temporary suspension of fine particles of a liquid (often an insecticide in oil) in the air with the same properties as a fog or mist. More commonly, a pressurised container (ATOMISER) with push-button nozzle, containing an insecticide (or other substance) in solution, with a liquefied gas as PROPELLENT (*q.v.*). See also NON-STOP AEROSOL.

Aggregation. (*phy*) The process of clumping together of insecticidal dusts and rodenticidal contact dusts, which may arise by their mis-use in damp locations, or by dusts being kept in store in water permeable containers, especially under conditions of high humidity. Aggregation may also be caused by vibration in transport and the physical interlocking together of different sized particles. As a result most commercially available dusts incorporate a low concentration of a 'free-flowing' or ANTICAKING AGENT (*q.v.*) which also improves application performance in dust gun equipment and 'puffer' packs of retail products.

Aggregation pheromone. (*ent*) A chemical produced by insects in small amounts, which keeps members of a population grouped together. Such substances have recently been demonstrated in cockroaches accounting for their typical aggregation in harbourages.

Agricultural Chemicals Approval Scheme. (*leg*) This scheme came into operation in the U.K. in June 1960 concerned then with products available in retail packs for horticultural and home garden use. It was extended in April 1970 to include products for use on grain in farm stores.

The purpose of the scheme is to enable users to select, and advisers to recommend, efficient and appropriate brands of agricultural chemicals (insecticides, fungicides and herbicides) and to discourage the use of unsatisfactory products.

The scheme is not concerned with safety requirements, but approval (i.e. endorsements of efficacy) cannot be given unless the product has first been considered under the Pesticides Safety Precautions Scheme. The A.C.A. scheme operates throughout the U.K., and participation in the scheme is voluntary. A certificate of Approval is granted to each product approved, and the container may so bear an identification mark (the Approval symbol and Registered number).

Air brick. (*bldg*) A specially designed brick or grille forming part of the coursing, inserted at intervals of about six feet, below the damp course level of buildings with suspended floors. Its function is to provide free ventilation of sub-floor spaces to reduce the moisture content of timber and susceptibility to rot. Some designs have holes sufficient to allow entry of small rodents, and when broken are not readily replaced, thus allowing the entry also of rats. Grilles of this type, often built into outside walls, provide the only source of ventilation of larders and pantries. Damage to these provides ready access for rodents to food, and in food manufacturing premises immediate replacement is essential to prevent rodent infestation.

Air curtain. (*equip*) Equipment emitting a 'curtain' of moving air, sideways or vertically downward, fitted to open doorways of buildings to prevent the entry of flying insects. Doorways of food manufacturing premises are difficult to proof against flies and wasps without impeding the movement of vehicles and fork lift trucks. Properly designed and competently commissioned, air curtains make a contribution to proofing where it would otherwise be impossible to stop the entry of flying insects.

Equipment is available for doorways up to 4 metres wide and 6 metres high; some combine the added advantage of supplying heat, as a warm air curtain in cold weather, preventing draughts of up to 13 km.p.h. from penetrating the doorway. See also FLY SCREENS and PLASTIC STRIP CURTAINS.

Alarm call chemical. (*chem*) A substance which causes agitation of individuals in a pest population when ingested, with the result that they exhibit certain symptoms, or cries of distress, causing the remainder of the population to disperse. See AVITROL.

Aldrin. (*chem*) An organochlorine insecticide introduced in 1948 under the trade name 'Octalene'. Aldrin is the name given to the 95% pure product. In the pure form it is known in Britain as HHDN. It is a contact insecticide closely related to, but more volatile than, dieldrin and as a technical product is a tan or dark brown solid. Aldrin is a persistent insecticide especially in soil, where it has fulfilled a major use (usually as an emulsion) in the control of subterranean termites. See TERMITE CONTROL. Additional uses include the control of fire ants (*Solenopsis* spp.) and occasionally as a contact dust for rats. Aldrin is readily absorbed through the skin. The acute oral LD_{50} (rat) is 70 mg/kg.

Alexandrine rat. (*zoo*) See RATTUS RATTUS.

Alimentary canal. (*zoo*) The gut; the food tube extending from mouth to

P.C.I.B.—B

anus, modified in various parts to fulfill specific functions (e.g. storage, digestion and absorption) which together with several organs forms the digestive system.

Allethrin. (*chem*) A synthetic pyrethroid with similar properties to pyrethrins, inferior in knockdown, but more persistent. First described in 1949; a pale yellow oil containing 75–95% of allethrin isomers. One of these is bioallethrin, the most active insecticidally, with good knockdown. As a contact insecticide, allethrin is as effective as pyrethrins against houseflies and is similarly often mixed with synergists (e.g. piperonyl butoxide) to enhance its action. For practical use, allethrin is often formulated in kerosene as a knockdown and flushing agent in fly sprays and aerosols. It is also used for the control of insect pests of stored and processed foods.

The acute oral LD_{50} (rat) of allethrin and bioallethrin is about 800 mg/kg.

Almond moth. (*ent*) See EPHESTIA CAUTELLA.

Alphachloralose. (*chem*) Glucochloral. A crystalline powder of low solubility in water, used in baits in the U.K. (under licence to the Ministry of Agriculture, Fisheries & Food) as a 'selective' method for the control of pest birds. It is used as a stupefying substance for the control of feral pigeons (at 1·5%) and sparrows (at 2%); those which succumb are killed by humane methods. PROTECTED SPECIES (*q.v.*) are allowed to recover and are released.

Alphachloralose (at 4%) was also introduced as a rodenticide ('Alphakil') for mouse control by Rentokil in 1966. It retards metabolism and lowers body temperature to a fatal degree in small mammals at 60°F (16°C) and below. Its mechanism of action offers a large measure of safety in use, causing rapid heat loss from a small mammal with a body of large surface area and low weight, compared with cats, dogs and man which have a smaller surface area to weight ratio and therefore a reduced speed of heat loss. Alphachloralose is rapidly metabolised and is hence non-cumulative.

Alphakil. (*p. prod*) See ALPHACHLORALOSE.

Alpha naphthyl thiourea. (*chem*) See ANTU.

Alphitobius diaperinus. (*ent*) Coleoptera: Tenebrionidae. The Lesser mealworm beetle. A pest of damp foods, grain and cereal products especially when mouldy, frequently increasing to astronomical numbers in the deep litter of poultry houses, especially those with earth floors. Control is required to prevent possible spread of poultry diseases between successive crops.

The adult (7 mm long) is oval, black, shiny. The larvae (10–15 mm) are yellow-brown and congregate around the food and water hoppers of poultry units. Pupae occur in the soil and fabric of the building. Minimum development period is about 40 days (25°C).

Aluminium phosphide. (*chem*) A highly insecticidal fumigant introduced in the early 1930s; stable when dry, but reacting with moist air to liberate the gas phosphine, which has a carbide-like odour and is spontaneously inflammable in air. Commercial formulations available are:

'*Detia GAS-EX-B*' (Freyberg Chemische Fabrik): crêpe paper bags containing 34g powder of 57% aluminium phosphide.

'*Delicia Gastoxin*' (DIA—Chemie): tablets (3g) of 57% a.i.

'*Celphos*' (Excel Ind., Bombay): tablets (3g).

'*Phostoxin*' (Degesch AG): tablets (3g) and pellets (0·6g) containing 55% aluminium phosphide, 40% ammonium carbamate (as a fire suppressant) and 5% aluminium oxide, formulated in paraffin.

The commercial products are used to fumigate a variety of commodities. Application rates vary with temperature: grain in silos (1 tablet per 1–3 tons), sacked goods (1 tablet per 2–4 m³). The moisture content of grain should not be less than 10%. Fumigation periods vary from 3–10 days. The residue is non-poisonous and is removed by screening the grain. Airing off may be required.

Tablets or pellets exposed on trays should be collected and removed from buildings at the end of treatment, likewise in the case of crêpe paper bags. Residues from aluminium phosphide preparations should not be heaped together as this constitutes a risk of fire or explosion.

The MAC value for aluminium phosphide in Germany is 0·1 ppm; USA 0·3 ppm. 2000 ppm of phosphine in air is rapidly lethal to man.

American cockroach. (*ent*) See PERIPLANETA AMERICANA.

American dog tick. (*zoo*) See DERMACENTOR VARIABILIS.

Amyl nitrite. (*chem*) See HYDROGEN CYANIDE.

Anagasta. (ent) See EPHESTIA.

Analytical methods. (*proc*) The means of separating, identifying and often measuring the components of a formulated pesticide for the purpose of 1) quality control in manufacture, 2) determination of pesticide residues in instances of contamination of perhaps food or water or 3) diagnosing the cause of death of a poisoned animal. The most commonly used methods are colorimetry and ultra violet and infra red spectrophotometry (for 1 above), chromatography in various forms and mass spectrometry, together with the previous techniques (for 2 and 3, above).

Anatomy. (*zoo*) The study of the structure of the body, principally the internal organs as shown by dissection. External anatomy, i.e. the detailed external appearance of the animal, is usually referred to as MORPHOLOGY (*q.v.*).

Angoumois grain moth. (*ent*) See SITOTROGA CEREALELLA.

Animal rooms. (*bldg*) Laboratories in which animals are reared or maintained for research purposes and which, by virtue of warm conditions, the presence of animal diets and frequent washing down, provide ideal conditions for the establishment of cockroach infestations.

Difficulties in cockroach control arise from the possible contamination of the environment with insecticides applied as dusts and sprays. Bait formulations, however, can provide complete eradication without effect on the test animals. The same applies to cockroach infestations in zoos, aquaria and pet shops.

Animals (Cruel Poisons) Act, 1962. (*leg*) An Act calling for the prohibition or restriction of poisons for the killing of animals where the use of such poisons would be cruel and unnecessary. The Act came into force in the U.K. as the Animals (Cruel Poisons) Regulations 1963 (Oct. 26th, 1963) prohibiting the use of certain rodenticides on grounds of cruelty (as interpreted by symptom expression in mammals). The regulations prohibit the use of phosphorus (elemental yellow phosphorus) and red squill (powder or extract of the red variety of *Urginea maritima*) as rodenticides, and the use of strychnine for destroying mammals with the exception of moles.

Anobiidae. (*ent*) The family of the Coleoptera containing stored product pests: LASIODERMA (*q.v.*) and STEGOBIUM (*q.v.*). Also some economically important wood-boring beetles, ANOBIUM and XESTOBIUM (*q.v.*).

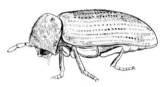

Anobium

Anobium punctatum. (*ent*) Coleoptera: Anobiidae. Common furniture beetle, House borer (New Zealand), often referred to imprecisely as 'woodworm'. An important pest of many timbers, notably the sapwood of softwood species, infesting structural timbers, flooring, joinery and furniture. The most prevalent wood-boring insect in buildings in Britain, living also outdoors in dead parts of trees, fencing and other structures. It occurs also in Europe and parts of Australia, New Zealand, South Africa and the United States.

Stegobium

Lasioderma

The adult (Fig. 2) is dark brown (2·5–5 mm long), covered with very short yellow hairs. A humped ('hooded') prothorax covers the head, when seen from above. It emerges from infested timber principally in June/July (U.K.) producing circular exit holes of 1–2 mm diameter. Does not feed; dies after 3–4 weeks.

Eggs (Fig. 3) are laid in crevices and on rough surfaces of timber and hatch in 2–4 weeks, the larva boring through the bottom of the egg, straight into the wood (cf. XESTOBIUM). The larva (6 mm long when fully grown), creamy white, curved and fleshy, lives in the wood for 3–5 years producing a network of tunnels in which a 'gritty' boredust of faecal pellets and wood fragments remains (Fig. 4). A pupal chamber is constructed just below the surface of the timber, the pupa (Fig. 5) hatching to adult after 3–8 weeks. Boredust beneath exit holes is evidence of an active infestation.

Spread occurs by flight of adults and the introduction of infested furniture and other household articles into buildings. About three-quarters of all buildings surveyed in Britain have been found to be infested.

Anopheles. (*ent*) Diptera: Culicidae. A large genus of mosquitoes; the eggs, each with a float, are laid singly on the surface of water (cf. CULEX). Larvae are supported horizontally below the surface of water by hairs. Pupae have conical respiratory trumpets. Adults rest with the abdomen at an angle with the surface. They breed in permanent areas of fresh water (ponds and lakes), the females selecting protected shore areas where the water is shallow. The adults are usually active only at night, spending the day in damp, dark protected places. They bite at dusk and just before dawn; flight is usually less than 1 mile from the hatching site. Some species transmit malaria. The principal species in Britain are *A. atroparvus, A. messeae, A. plumbeus* and *A. claviger*. Species of importance in the U.S. include *A. punctipennis*: the most widely distributed Anopheline mosquito, not a carrier of malaria, breeding in trapped water (in containers, swamps, bogs). Also *A. quadrimaculatus* the common malarial mosquito of the eastern and southern United States. Egg laying is continuous during warm weather, the life cycle being completed in 8–14 days.

Anoplura. (*ent*) The Order of insects containing the sucking lice; external parasites of mammals, the mouthparts modified for piercing the skin and taking blood. Small flattened wingless insects, the head much narrower than the thorax. Examples of pest species are *Pediculus humanis*, the Body louse and *Linognathus setosus*, the Dog sucking louse.

Antagonist. (*chem*) A chemical substance which reduces the action of another and *vice versa*, as when one drug with a certain physiological action is given simultaneously with, or soon after, another which has an opposite action. The principle of antidotal treatment. See ANTIDOTE.

Antenna. (*ent*) One of a pair of mobile, segmented appendages articulated with the head, in front of, or between, the eyes of insects. They vary widely in form and size having the sensory functions of taste (Hymenoptera) and smell (Blattaria and Lepidoptera). Because of the waving movement of the antennae of some insects (e.g. cockroaches), apparently sensing their surroundings, these organs are frequently referred to as 'feelers'.

Anthrenus verbasci. (*ent*) Coleoptera: Dermestidae. The Varied carpet beetle (Fig. 6), an increasingly important pest of domestic properties in suburban areas of S. and S.E. England, infesting carpets, blankets and furnishings. Also a pest of dried insect collections and most products of animal origin.

The larvae ('woolly bears') are small (4–5 mm), brown and hairy with bunches of golden, spear-headed hairs either side of the rear segments. The larvae tend to roll up into tiny golden balls when disturbed and can survive long periods of starvation. The pupa is formed within the last larval skin. Adults normally emerge in May–June, are oval, strongly convex with white, black and yellow scales on the head, pronotum and elytra. These feed outdoors on the flower heads of Hogweed and Spiraea.

A. verbasci is a frequent resident of birds' nests subsisting on feathers; it gets into premises *via* the roof void, pipe lagging, airing cupboards and thence into bedrooms and ground floor rooms where stored woollens and the edges of carpets are particularly susceptible to attack (Fig. 1).

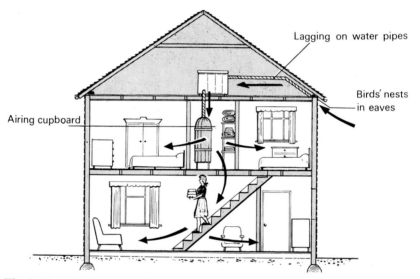

Fig. 1   Spread of beetles which attack textiles in the home, showing movement from birds' nests to airing cupboards and then to rooms.

Anticaking agent. (*chem*) A substance incorporated into an insecticidal or rodenticidal dust to maintain free-flowing properties. Examples are fumed silicas (Cabosil), tricalcium phosphate, magnesium oxide, magnesium carbonate and magnesium stearate, usually incorporated at about 1%.

Anticoagulants. (*chem*) Chemicals which inhibit the clotting of blood, used widely as rodenticides in cereal and liquid baits and as contact dusts. The rodent bleeds to death through internal and external haemorrhages.

In a normal animal, clotting of the blood is brought about by a soluble substance (fibrinogen) being changed into insoluble fibrin, the 'strands' of fibrin entangling the blood cells to form a clot. This change can only take place in the presence of an enzyme (prothrombin) continuously released from the liver into the blood. In the poisoned animal, production of this enzyme is inhibited, with the result that the blood fails to clot.

Anticoagulant baits must usually be eaten over a period of several days (usually 5–14) to exert their effect: mice take longer to kill than rats. Two types of anticoagulant are recognised: hydroxy coumarin compounds and indanedione compounds. They rarely lead to BAIT SHYNESS (*q.v.*) at the concentrations used because their effect on the body is delayed, and rats and mice do not associate symptoms of poisoning with the bait. By the time symptoms appear, a lethal dose has been acquired.

Generally regarded as 'safe' rodenticides, but not uncommonly, acute doses may be taken by domestic animals resulting in death. The most widely used anticoagulant is WARFARIN (*q.v.*). Vitamin K is an antidote for poisoning by anticoagulants. The hazard of secondary poisoning is low.

Antidote. (*tox*) A chemical substance which when administered counteracts poisoning, either by chemically reacting with the poison, changing it into a harmless substance, or by setting up an action in the body opposite to that of the poison. One of the most widely known antidotes in pest control is Vitamin K, the antidote for warfarin poisoning. In some countries it is a requirement of the labelling of pesticides that information on antidotes be stated. See PRALIDOXIME CHLORIDE; also HYDROGEN CYANIDE.

Antimetabolite. (*chem*) A chemical that interferes with or antagonises normal metabolic processes, e.g. an insect's ability to metabolise food, so that it starves to death. Used successfully in protecting fabrics. In experimental stage of development. Usually of low toxicity to man and animals.

Antimony potassium tartrate. (*tox*) Tartar emetic. A white crystalline compound, once popular as a medicine, poisonous in large doses, largely replaced by safer drugs. See EMETIC.

Antiseptic. (*chem*) A substance which destroys or inhibits the growth of bacteria; a proprietary product should be carried in the pest controller's first aid kit for the treatment of cuts and dermal abrasions.

Ants. (*ent*) See FORMICIDAE.

Antu. (*chem*) Alpha naphthyl thiourea, alpha naphthyl thiocarbamide. An acute rodenticide, little used today, effective against *Rattus norvegicus* (more against adults than young) but with little value against *Rattus rattus* and *Mus musculus*. The technical product is a grey powder, stable on exposure to air, used in baits at 2–3%, and as a 20% contact dust.

Aversion and tolerance to Antu lasting up to 6 months is rapidly developed in rats through repeated takes of sublethal doses. It has been withdrawn from use in the U.K. because of the carcinogenic properties of beta naphthylamine, possibly present as an impurity. The acute oral LD_{50} (rat) is 6–8 mg/kg. It is not absorbed through the skin.

Aphodius rufipes. (*ent*) See DUNG BEETLE.

Apis mellifera. (*ent*) Hymenoptera: Apidae. The Honey bee, for which no description is necessary, occasionally a serious pest when nests are established in chimneys, under tiles or in wall voids of homes. The nest is made of wax cells produced by the workers; some are used by the developing immature stages and others for the storage of honey. The primary concern of the home owner is that he may be stung, especially when bees find their way into the house: usually the barbed sting and venom sac remain in the flesh as the bee is brushed away. A secondary concern is possible structural damage as a result of nest building activities, and staining of ceilings and walls due to leakage of honey.

In areas near hives care is necessary to prevent contamination with insecticides, especially if treatments are being carried out for wasp control involving insecticidal baits.

Apodemus sylvaticus. (*zoo*) The Long-tailed field mouse, or Wood mouse, easily confused with *Mus musculus*, but for the white underside of the body, larger ears and eyes, and longer tail (Fig. 7). Not a pest of food manufacturing premises or warehouses, but occasionally sheltering indoors in the autumn, frequently entering domestic properties, especially where apples are stored.

Approval of pesticides. (*leg*) The process of bringing pesticides under legislative control (voluntary at present in the U.K.), resulting in authorisation of the 'notifier' (a manufacturer or servicing company) to sell and/or use a pesticide. The authorising board is entitled to request the notifier to provide all the necessary information as to the chemical composition, toxicity and proposed use of the pesticide, and information on such technical subjects as methods of residue analysis of the active ingredient. See PESTICIDES SAFETY PRECAUTIONS SCHEME.

In some countries the approval of pesticides operates as a form of registration or licensing and involves tests of efficacy for the purpose intended, as well as approval on grounds of safety in use. Because of the Federal structure of the U.S.A., control of application of pesticides falls within the competence of the individual States. Federal registration of a given product does not make State registration unnecessary where this is required by law.

Arachnida. (*zoo*) The class of the Arthropoda containing mites, (many of which infest stored foods), spiders and scorpions (household pests) and ticks (parasites of animals). All with eight legs and the body undivided, or divided into two parts (cephalothorax and abdomen). See ACARI (mites and ticks) and ARANEAE (spiders).

Araneae. (*zoo*) The family of the Arachnida containing the Spiders. Animals of characteristic appearance, with the body divided into 1) a head-thorax (cephalothorax or prosoma) with 6–8 eyes, mouthparts and 4 pairs of legs, and 2) abdomen (opisthosoma) with silk-spinning spinnerets. Carnivorous, some species using a web as a trapping device for insect prey.

Objectionable in the U.K. solely by their presence (see TEGENARIA) and the spinning of webs; a number of venomous species occur in the U.S. causing 'necrotic spider bite'. See LOXOSCELES.

Architrave. (*bldg*) A covering, usually of timber, around the frame of a door, or occasionally a window, hiding the join of the plaster with the timber work. In practice the architrave is rarely flush with both surfaces, providing crevices favoured as harbourages by cockroaches and other insects.

Arcton. (*p. prod*) See PROPELLENT.

Argentine ant. (*ent*) See IRIDOMYRMEX HUMILIS.

Arprocarb. (*chem*) See PROPOXUR.

Arsenic trioxide. (*chem*) White arsenic, arsenious oxide. An inorganic, acute rodenticide, little used today; a heavy white, odourless and tasteless powder combined into baits at 3%; more effective when finely ground. Inexpensive and relatively slow acting. Acute oral LD_{50} (rat) 25 mg/kg. Also once used in pastes for ant control. Still used as a 30% dust in some countries for application to termite galleries for transfer by the insects back to the nest.

Arthropoda. (*zoo*) The largest phylum of the animal kingdom whose members have many characters in common, including a segmented body, chitinised exoskeleton, and paired jointed appendages, modified according to function, carried on a variable number of segments.

The classes of this phylum which include pest species are the Insecta (insects), Arachnida (scorpions, spiders, mites, ticks, etc.), Diplopoda (millipedes) and Chilopoda (centipedes). The only additional class of any size is the Crustacea (lobsters, shrimps, crabs and barnacles) which includes the woodlice.

Artificial respiration. (*proc*) The stimulation of respiration in someone whose breathing has stopped. Also known as resuscitation. This requirement in pest control is more likely after electric shock from ill-maintained electrical equipment than from accidental poisoning by pesticides.

If breathing stops, permanent damage to brain tissues begins within four minutes. Artificial respiration should be started immediately. Two methods are recommended. 1) Mouth to mouth (kiss of life); it gives the greatest inflation of the lungs and oxygenation of the blood; the degree of inflation can be controlled by watching the chest; it is the least tiring of methods. 2) The Holger-Nielsen method; the best of other methods, although Schafer's method is often recommended.

It is recommended that all engaged in pest control (especially in fumigation) should be thoroughly acquainted with the techniques of resuscitation. They are not difficult to learn once the principle is understood.

Atomiser. (*equip*) A technically incorrect name for a piece of equipment, mechanical, thermal, or using compressed gas, to produce a mist or fog of minute droplets: used for the insecticidal treatment of spaces (against flying insects) and occasionally for disinfection. Such equipment is rarely suitable for the control of crawling insects, or the immature stages of flies or moths, unless the 'atomised' spray is directed into harbourages. See AEROSOL.

Atropine sulphate. (*chem*) An antidote for the treatment of poisoning by organophosphorus insecticides, relieving many of the distressing symptoms, reducing heart block and drying secretions of the respiratory tract. The dose for usual cases of poisoning is 1–2 mg given intramuscularly to a maximum of 25–50 mg in a day. Effects of injection become evident in 1–4 minutes and maximal within 8 minutes. Patients should remain under medical supervision for at least 24 hours. See also PRALIDOXIME CHLORIDE.

Attagenus. (*ent*) Coleoptera: Dermestidae. *Attagenus pellio*, the Fur beetle, is a common pest of homes; infestations also occasionally become established in clothing stores. The materials most often attacked are, hair, furs, skins, feathers, woollen fabrics, carpets and upholstery. Considerable damage may occur before the pest is detected. *A. pellio* is common in birds' nests, and may also occasionally infest stored grain where it feeds principally on dead insects. The larvae (also called 'woolly bears'—see *Anthrenus verbasci*) are elongate, with a distinctive tuft of very long golden hairs on the last segment; larvae are about 6 mm long; cast skins are often a feature of an infestation. Adults are oval (4–6 mm) dark red-brown to black with a characteristic patch of white hairs in the centre of each elytron. Life cycle and biology are similar to ANTHRENUS VERBASCI (*q.v.*).

In the U.S.A. and Asia, *Attagenus piceus*, the Black carpet beetle, wholly dark brown to black, assumes greater importance than *A. pellio*, the more abundant species in Britain.

Attractant. (*chem*) A substance, usually volatile, often incorporated into a bait with the object of drawing the pest (rodent or insect) towards it from a distance. The use of an attractant aims at stimulating the olfactory senses of the pest; it is not to be confused with a bait 'additive' used to stimulate the chemo-receptors of the mouth with the purpose of encouraging the pest to feed more readily.

Many insects are known to emit minute quantities of sex attractant which bring the sexes together for mating (see PHEROMONE). Ultra violet light (a physical attractant) has been exploited in various electric devices for attracting flying insects and killing by electrocution.

Australian cockroach. (*ent*) See PERIPLANETA AUSTRALASIAE.

Australian spider beetle. (*ent*) See PTINUS TECTUS.

Automatic insecticide dispenser. (*equip*) Equipment designed to provide a continuous or intermittent discharge of insecticide into the air; e.g. THERMAL VAPORISING UNIT (*q.v.*) emitting lindane vapour, SLOW RELEASE STRIP (*q.v.*) emitting dichlorvos, and mechanical equipment delivering measured doses of an aerosol (synergised pyrethrum or synthetic pyrethroid) at prescribed intervals.

Autumn fly. (*ent*) See MUSCA AUTUMNALIS.

Avicide. (*chem*) A substance which kills birds.

Avitrol. (*p. prod*) An alarm call chemical (4-aminopyridine) formulated in
baits at 0·5–1 % to induce sparrows, starlings, pigeons and gulls to leave
an area by producing flock disturbing symptoms. Avitrol temporarily
affects a bird's ability to fly causing it to emit cries and give other signs
of distress which scare away other birds in the flock. The technical
product is an off-white solid; toxicity to birds is high, LD_{50} sparrow
4 mg/kg, gulls 8 mg/kg. Available as formulated baits and concentrates.
Acute oral LD_{50} (rat) 32 mg/kg.

Bacillus. (*dis*) A genus of rod-shaped, spore-forming bacteria which includes a number of physiologically important species. *Bacillus thuringiensis* causes 'wilt disease' among larvae of *Ephestia elutella* and is now commercially available as a method of biological control.

Spores have also been incorporated into poultry feeds to prevent flies breeding on chicken dung. Formulation available: spore suspension.

Bacon beetle. (*ent*) See DERMESTES LARDARIUS.

Bacteria. (*dis*) A group of microscopic organisms consisting of one or many cells; varying in shape, often rod-like, more or less spiral, filamentous or in some forms spherical; lacking chlorophyll. The majority multiply by simple division, but other forms of asexual, as well as sexual reproduction, are known. Two major groups are recognised: gram-positive and gram-negative, reacting differently to staining procedures.

Bacteria are ubiquitous, occurring under favourable conditions in massive numbers; many are concerned in the breakdown of plant and animal tissues. Those of importance in pest control are the bacteria transmitted by rodents, insects and birds, causing serious diseases in man (Fig. 11). A working knowledge of bacteriology is important to those involved in sanitation and public health.

Bactericide. (*chem*) A substance which kills bacteria.

Bacteriostat. (*chem*) A substance which inhibits the growth of bacteria but does not kill them (cf. BACTERICIDE).

Bait. (*proc*) A formulation of a pesticide with an attractive food, occasionally with water as a drink; the commonest method of control of rodents, which for success requires a knowledge of rodent behaviour and preferences for bait materials. For safety in use, baits must be placed with care, inaccessible to children and domestic animals (Fig. 8).

Bait bases most commonly used are cereals, fruits, fish and meats, the bait usually placed in containers to avoid spillage and possible contamination of food and water. Liquid baits are effective in hot dry locations; variation in bait base is often necessary to achieve the desired results. PRE-BAITING (*q.v.*) increases the effectiveness of acute rodenticides. See also TEST BAITING. LIVER BAITING (*q.v.*) as a test bait, is a valuable first step in the control of certain indoor ants.

Baits incorporating STOMACH POISONS are used in insect control; as supplementary measures to dusts and sprays for the control of cockroaches and ants (see CHLORDECONE); for fly control (see TRICHLORFON) and for wasps (see CARBARYL). The objective in all cases is to attract pests to the baits in preference to other foods, to kill them directly, or in the control of ants and wasps for the transfer of insecticide back to the nest.

Baiting is also used for control of birds (see ALPHACHLORALOSE, AVITROL) and as an adjunct to TRAPPING (q.v.).

Bait block. (*chem*) Rat block. A rodenticide formulated in a food base and wax (sometimes with flavouring agents) for use in places where dampness and moulds may cause rapid deterioration of dry cereal baits.

Bait box. (*equip*) A container, usually for rodenticidal baits, often made of wood with a hinged lid to contain a large amount of bait (cf. BAIT TRAY) as a permanent bait station in an outdoor location. The box usually has small areas cut out from each end to allow rodents free entry and exit. Some are fitted with internal baffles to prevent bait spillage. The purpose of the bait box is to protect baits from the weather and 'takes' by other animals. See also BAIT TUNNEL.

Bait shyness. (*zoo*) The characteristic behaviour of rodents, which have suffered or recovered from rodenticidal poisoning, of avoiding those baits (usually containing acute poisons) which caused the symptoms of distress. A condition brought about by rodents associating distress with their most recent food source. An important consideration in the effective use of acute rodenticides, often overcome by first PRE-BAITING with unpoisoned food, thus encouraging a lethal 'take' at the first feed.

Bait station. (*equip*) See BAIT BOX, BAIT TRAY, BAIT TUNNEL.

Bait tray. (*equip*) A small cardboard or plastic tray used in buildings to contain rodenticidal bait for the control of rats and mice. Its purpose is: to prevent the contamination of surfaces and often foodstuffs by rodenticide; to keep the bait from direct contact with wet surfaces; to establish a bait station (a permanent baiting point) which can be inspected frequently to assess, by the presence of droppings and by the amount of bait eaten, whether or not the infestation is being controlled. Also to provide available rodenticide for the control of rodents which may subsequently gain entry.

Bait tunnel. (*equip*) An elongate container for rodenticidal baits, open at both ends to protect the bait from being taken by other animals (Fig. 9). Of particular value outdoors where acute poisons are used; earthenware land drains (Fig. 8) and other short ($\frac{1}{2}$ m) sections of pipe serve this purpose. Also occasionally made from cardboard, to prevent baits indoors in manufacturing industries from becoming covered by dust.

Balaustium murorum. (*zoo*) Acari: Erythraeidae. Red spider mite. An intruder, causing annoyance but no damage, often confused with BRYOBIA (q.v.) the Gooseberry red spider mite. Of similar size, but brighter red and running at great speed in a haphazard manner. Feeds exclusively on pollen. Properties are invaded in spring and autumn, the mites entering fortuitously through cracks in windows. Eggs are laid in crevices in walls and in the soil.

Balustrade. (*bldg*) The infilling of small pillars to a raised parapet, rail or coping around the edge of a roof. Such a structure often provides a day perch for feral pigeons and requires treatment with a tactile repellent.

Barge. (*ship*) Lighter. A flat-bottomed vessel, usually without sails or screws, used for carrying goods in bulk on rivers and canals not navigable by larger craft. Barges are also frequently used for transhipping commodities at port between the supply vessel and the dock side. They provide a convenient means of fumigating commodities in small parcels under tarpaulins. Two types of lighters are in general use; 1) 'hatched craft' which have a centre ridge beam, removable hatch boards and fitted with cleats and wedges for battening down tarpaulin covers and 2) 'open craft' without a centre ridge and usually no provision for battening down. The fumigation procedure has to be modified accordingly.

Barium carbonate. (*chem*) A very old rat poison, relatively inefficient, chemically inert and not highly toxic to man. Barium carbonate is a dense white powder, odourless and non-volatile, and only slightly soluble in water. Rat baits are prepared from 1 part barium carbonate with 4 parts food base moistened with milk or water. Acute oral LD_{50} (rat) 630–750 mg/kg.

Basudin. (*p. prod*) See DIAZINON.

Bats. (*zoo*) See CHIROPTERA.

Baygon. (*p. prod*) See PROPOXUR.

Baytex. (*p. prod*) See FENTHION.

Bean beetle/weevil. (*ent*) See ACANTHOSCELIDES OBTECTUS.

Bedbug. (*ent*) See CIMEX LECTULARIUS.

Beetles. (*ent*) See COLEOPTERA.

Behaviour. (*zoo*) The response of an organism to the stimuli of its environment, together with inherent characteristics of the species (e.g. circadian rhythm). A knowledge of animal behaviour is often of value in pest control, e.g. in anticipating the responses of rats and mice in their 'new object reaction' to the positioning of rodenticidal baits.

Benching. (*bldg*) The raised platform at the bottom of a sewer entrance (manhole) used for keeping rat baits above the level of the effluent. When there is no benching, bait may be suspended above the effluent in a muslin bag. See also SEWER TREATMENT.

Bendiocarb. (*chem*) An experimental carbamate made available by Fisons as test samples in 1971. The technical product is a white crystalline solid with low solubility in water and not readily soluble in organic solvents. 'Ficam 80W' is an 80% wettable powder with claims for rapid and high activity against cockroaches and good effect on other insects of importance in public health, including ants. Recommended applications are 0·25%–0·5% spray to run off; good residual activity is claimed with lack of smell and irritant properties. The acute oral LD_{50} (rat) is 60–120 mg/kg. The compound is rapidly metabolised and excreted by mammals.

Benzene. (*chem*) A liquid distillate of coal tar of high volatility; a good solvent for most pesticides but odorous, and little used because of its high mammalian toxicity and inflammability (more so than ODOURLESS KEROSENE *q.v.*).

BHC. (*chem*) Benzenehexachloride. The common name for the six mixed ISOMERS (*q.v.*) of hexachlorocyclohexane, an organochlorine compound with insecticidal properties discovered in the early 1940s by ICI Limited. Only one of the isomers (see GAMMA-BHC) has useful insecticidal properties and it is on this that formulations of BHC for industrial and domestic use are based.

BHC is an off-white to brown powder, having a heavy, persistent and musty disagreeable odour. Its main use is in agricultural insect control where these undesirable properties are of little consequence; it can however seriously taint certain crops (e.g. black currants). It is a stomach poison and contact insecticide with some fumigant action. The mammalian toxicities of the isomers are different.

Bioallethrin. (*chem*) See ALLETHRIN.

Bio-assay. (*proc*) A test method using living organisms to evaluate the biological activity of a substance, usually a new compound, with the object of 'screening' the substance for its usefulness as a pesticide. Occasionally also used as a method of quality control in pesticide manufacture.

Biochemistry. (*chem*) The science of the chemistry of living organisms.

Biocide. (*chem*) A substance with activity against biological organisms.

Biological control. (*proc*) The control of pests by other biological organisms. To some extent biological control exists in all animal populations by the natural predation or parasitism of one species upon another and the competition between individuals for space. In recent years, however, attention has been drawn to the merits of control methods using bacteria and viruses, as alternatives to chemicals, without undesirable effects on the environment.

Biological control also offers the potential of specificity of action without effects on desirable species.

During the period 1928 to 1958 *Salmonella enteritidis* (var. *Danysz*) was used in the U.K. by the British Ratin Co. (now Rentokil) as a bait preparation for rodent control. This organism was isolated by Danysz in 1900 from an epidemic of mouse typhoid in field mice and has always been considered to be predominantly a parasite of rodents.

Myxomatosis is an example of biological control (self spreading) which has proved highly effective in drastically reducing rabbit populations. See also BACILLUS.

Bioresmethrin. (*chem*) See RESMETHRIN.

Bird control. (*proc*) The application of legally approved chemical and physical methods (see BIRD LAWS), to reduce or eliminate a local or area problem created by pest birds, normally *Columba livia, Passer domesticus* and *Sturnus vulgaris*.

There are nine good reasons for bird control:

1) Prevention of defacement of buildings by droppings, which accelerate the deterioration of masonry.
2) Reduction in cleaning costs in restoring the appearance of buildings.
3) Reduction in maintenance costs: nests and droppings retard water run-off leading to timber decay and broken renderings.
4) Prevention of damage in manufacturing industries: droppings mar finished products in loading bays rendering them unfit for sale.
5) Prevention of food contamination: droppings, regurgitated pellets, feathers and nesting material are common contaminants of grain.
6) Reduction in disease potential: the exact role of birds in disease transmission is not well known but Ornithosis can be transmitted from pigeons to man in the U.S.
7) The elimination of noise and smell.
8) Removal of sources of insect and mite infestation provided by nests, their excreta and the birds themselves.
9) Improvement in the safety of air traffic at major airports.

Where direct killing is prohibited, control of local populations relies on the use of TACTILE REPELLENTS (*q.v.*), stupefying substances (see ALPHACHLORALOSE), alarm call chemicals (see AVITROL) and BIRD SCARING DEVICES (*q.v.*). Area control relies on the use of TRAPPING (*q.v.*) and sterilants (see ORNITROL).

Bird distress call recordings. (*equip*) See BIRD SCARING DEVICES.

Bird fleas. (*ent*) See CERATOPHYLLUS.

Bird Laws. (*leg*) Legislative restrictions on methods of control of pest birds, covered in the U.K. by the Protection of Birds Act, 1954 and 1967 which among other things, prohibit the use of chemicals for direct killing. Control methods in the United States, must conform with Federal Bird Protection Laws (which do not afford protection to feral pigeons, house sparrows and starlings) but there are also State and Local laws which involve licensing and the regulation of techniques used in bird control.

Bird lice. (*ent*) See MALLOPHAGA.

Bird proofing. (*proc*) Prevention of entry of pest birds (usually pigeons and sparrows) into buildings by the use of galvanised chicken wire, expanded metal (aluminium) or netting (polythene coated or tarred nylon); to supplement the treatment of buildings with tactile repellents and stupefying baits. Structures often requiring proofing are open windows, louvres, niches around statuary, ventilation points and open eaves. Nets for pigeons should be $1\frac{3}{4}$ in. mesh and for sparrows $\frac{1}{2}$ in. mesh (Fig. 10).

Bird scaring devices. (*equip*) For frightening pest birds from an area where they roost or cause hazard (e.g. to aircraft, see LARUS), including loud noises, flashing lights and replayed recordings of bird distress calls. Of limited value unless used sparingly and intermittently: most birds learn to ignore them by repeated experience. See also TACTILE REPELLENT and AVITROL.

Biscuit beetle. (*ent*) See STEGOBIUM PANICEUM.

Biting lice. (*ent*) See MALLOPHAGA.

Stages in the development of
Anobium punctatum.
Fig. 2 (*above*) adult;
Fig. 3 (*left*) eggs on the end grain
of timber;
Fig. 4 (*below*) larva in a tunnel in
plywood;
Fig. 5 (*right*) pupa in pupal
chamber.

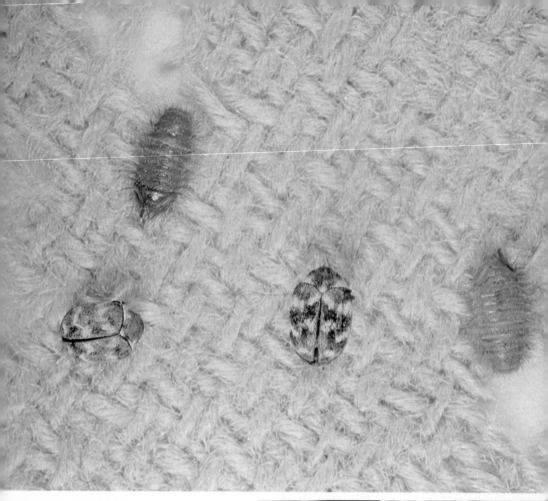

Fig. 6 (*above*) Adults and larvae ("woolly bears") of *Anthrenus verbasci* on damaged fabric.

Fig. 7 (*right*) *Apodemus sylvaticus* gnawing an acorn. Note the characteristic large eyes and ears and the white underside. An outdoor mouse occasionally entering buildings. (Kinns).

Fig. 8 (*right*) Use of a drain pipe to protect rodent baits from the weather and from being eaten by domestic animals. The structure of this building provides ideal harbourage for rats with plenty of cover in the vicinity.

Fig. 9 (*below*) Use of cardboard tunnels to provide permanent outdoor bait stations for rodent control.

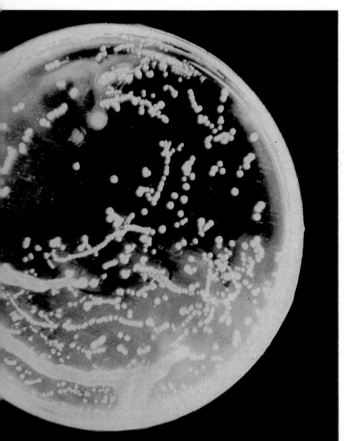

Fig. 10 (*above*) Use of nylon netting to proof a building against the entry of feral pigeons, supplementary to a treatment with tactile repellent (see Fig. 97)

Fig. 11 (*left*) An agar plate (nutrient gel) over which a cockroach has walked, inoculating the plate with pathogenic bacteria. After incubation the bacteria grow to produce large, easily visible colonies.

Fig. 12 (*right*) Female *Blatta orientalis* with oötheca. The wings of the female are shorter than in the male.

Fig. 13 (*below*) Female *Blattella germanica* on tomato. The two dark bands on the pronotum are clearly visible. Fruits and vegetables provide cockroaches with an essential supply of water.

Castes of termites:
Fig. 14 (*left*) soldier heavily armoured, with enlarged head and mandibles;
Fig. 15 (*below*) alate reproductive.

Fig. 16 (*below*) workers in infested timber; note their pale colour and the accumulation of faecal pellets, packed into disused galleries.

Fig. 17 (*above*) A typical roosting site of *Columba livia*, occupied at dusk, the birds preening themselves prior to settling for the night.

Fig. 18 (*below*) A heated, modern aviary where chlordecone bait may be used most effectively and safely against cockroaches.

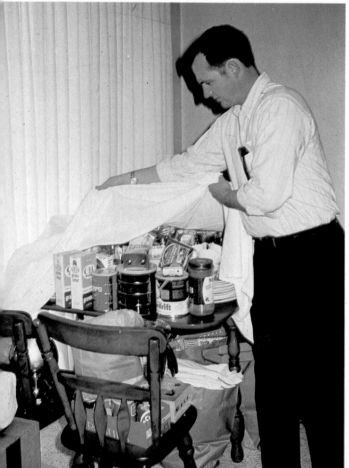

Fig. 19 (*above*) Application of contact dust to a rat burrow outdoors. Suitable for use only in dry locations.

Fig. 20 (*left*) Cockroach control often requires the treatment of food cupboards. Commodities should be removed and protected from contamination by spray mist.

Black beetle. (*ent*) See BLATTA ORIENTALIS.

Black carpet beetle. (*ent*) See ATTAGENUS PICEUS.

Black garden ant. (*ent*) See LASIUS NIGER.

Black rat. (*zoo*) See RATTUS RATTUS.

Black widow spider. (*zoo*) See LATRODECTUS MACTANS.

Blaps mucronata. (*ent*) Coleoptera: Tenebrionidae. Churchyard or Cellar beetle. Adult entirely black (22 mm long), slow and cumbersome in movement; occurs in stables, cellars and other dark, damp locations (including coal mines). A nocturnal, casual intruder, the larvae feeding on vegetable matter.

Blatta orientalis. (*ent*) Dictyoptera: Blattidae. The Oriental cockroach, Black beetle, or Shad roach. Thought to have originated from North Africa, now widely distributed by trade to almost all temperate regions of the world, preferring less humid and cooler conditions (22–27°C) than *Blattella germanica*, and *Periplaneta* spp. A major pest of buildings, notably in cellars, basements, ducts, boiler rooms, kitchens, bakeries, toilets and bars.

The major domiciliary cockroach in Britain, a potential carrier of pathogenic bacteria when associated with food. Occasionally able to survive the winter outdoors, notably in refuse heaps.

The adult Oriental cockroach is large (20–24 mm long) dark red-brown to black, the female with much reduced wings and tegmina, neither sex able to fly (Fig. 12). Not often seen during the day, less active than *B. germanica*. The oötheca (10 mm long) is brown-black and deposited by the female soon after formation. Oöthecae are produced at about 10-day intervals and contain up to 18 eggs. There are 7–10 nymphal stages, development to adult taking 6–12 months in heated buildings. The adult life span varies from 2 months under warm conditions to 9 months in cool conditions.

Blattaria. (*ent*) The suborder of the Dictyoptera containing the cockroaches, or roaches. They have the following general appearance: body oval and flattened. Head when at rest nearly horizontal and bent under the thorax, with the mouth projecting backwards between the bases of the first pair of legs; almost concealed from above by an enlarged pronotum. The whip-like antennae inserted just below the middle of the eyes, are composed of very many short segments and are often longer than the body. The mouthparts serve a biting, chewing and licking function; the mandibles are strong and toothed, the palps of the maxillae and labium have 5 and 3 segments respectively. The ocelli are often represented by two thin areas of cuticle. All three pairs of legs of cockroaches are long and well adapted for running. The gizzard (proventriculus) contains an armature for masticating food.

Most cockroaches are drab brown in appearance and predominantly tropical. Some, (the pest species) have occupied the artificial environments (buildings, sewers, ships) created by man, who has been largely responsible for their spread.

See BLATTA ORIENTALIS, BLATTELLA GERMANICA, B. VAGA, PARCO-
BLATTA SPP., PERIPLANETA AMERICANA, P. AUSTRALASIAE, P. BRUNNEA,
P. FULIGINOSA, P. JAPONICA, SUPELLA LONGIPALPA.

Blattella germanica. (*ent*) Dictyoptera: Blattidae. The German cockroach
or steamfly. Thought to have originated in N.E. Africa, and now distributed by commerce to virtually all parts of the world. It prefers warm
moist conditions (25–30°C) and has consequently become a common
pest of heated buildings, notably in kitchens, larders and restaurants.
It is the most prevalent cockroach in the galleys, store rooms and
accommodation of modern ships. In the southern States of America,
and other warm countries, the German cockroach occurs outdoors but
this is rare in Britain.

The adult German cockroach is small (10–15 mm) buff coloured with
two distinct dark parallel bands running the length of the pronotum
(Fig. 13). The wings are as long as the body in both sexes. It is generally
more active than the Oriental cockroach and more readily seen during
the day. The oötheca (8 mm long) is chestnut brown segmented externally (about 18 segments) and is carried by the female until the eggs
hatch. Oöthecae are produced at about 3–4 week intervals and contain
35–40 eggs. There are usually 6 nymphal stages, development to adult
taking 6–7 weeks at 30°C. The adult life span is 4–5 months.

Blattella vaga. (*ent*) Dictyoptera: Blattidae. The Field cockroach. Similar
in size and colouring to *B. germanica*, but differing in the presence of a
black line between the eyes and mouth of *B. vaga*; most common in the
irrigated regions of S. Arizona, occurring under stones and feeding on
plant debris. In dry seasons the Field cockroach may enter homes in
large numbers where it may remain and breed; usually, however, a casual
intruder.

Blattanex. (*p. prod*) See PROPOXUR.

Blooming. (*chem*) Of an insecticide, the formation of minute crystals of the
active chemical on a surface following application by spraying; caused
by rapid loss of solvent by evaporation, with the result that the insecticide has insufficient solvent for it to be absorbed in solution into the
surface. Also, a characteristic of insecticidal lacquers containing
insecticides (as solids), the bloom of crystals being regenerated on the
surface through the migration of active chemical in the lacquer film. See
also PENTACHLOROPHENOL.

Blow fly or **Blue bottle fly.** (*ent*) See CALLIPHORA ERYTHROCEPHALA.

Body louse. (*ent*) See ANOPLURA.

Bolting cloth beetle. (*ent*) See TENEBROIDES MAURITANICUS.

Booklice. (*ent*) See PSOCOPTERA.

Borax. (*chem*) The common name for sodium tetraborate decahydrate
(= sodium biborate = sodium pyroborate), a white crystalline solid,
efflorescent in dry air, and once used alone, or in mixtures, as a powder
for domestic cockroach control (see also BORIC ACID). Soluble in water
(5 g/100 ml at 20°C), long used as a mild antiseptic and more recently

as a component of fire retardant formulations. Also used as a corrosion inhibitor for ferrous metals, a herbicide (because of its strong phyto-toxicity) and as an additive to sodium chlorate to reduce fire risk. The acute oral LD_{50} (rat) is 2·7–5·1 g/kg. The lethal dose to infants is 5–6 g.

Boredust. (*ent*) See FRASS.

Boric acid. (*chem*) Orthoboric acid, Boracic acid. An inorganic insecticide used for cockroach control for many years before the introduction of synthetic organic compounds; more effective and more rapid in action than BORAX. Interest has been renewed in boric acid because of its non-repellent properties compared with more recently introduced insecticides. Boric acid admixed with cereals and sugar as a bait is no more effective than boric acid alone as a dust.

A white crystalline solid, slightly soluble in water (6%), of extremely low volatility. A slow acting stomach poison and of long insecticidal persistence.

Of low oral and dermal toxicity; acute oral LD_{50} (rat) is 2700 mg/kg; however, use at high concentrations, (approaching 100%), does not make boric acid less hazardous than more toxic insecticides used at lower concentrations. A 'free-flowing' agent is usually incorporated to prevent particle aggregation when boric acid is used as a dust. Past use has involved admixture with sodium fluoride (25%) and/or pyrethrins (up to 3%).

Botanical insecticides. (*chem*) Insecticides made from extracts of certain parts of plants, including nicotine and some of the safest insecticides (e.g. pyrethrum, derris). Synthetic compounds simulating the action of some plant extracts have recently been introduced. See SYNTHETIC PYRETHROIDS.

Boxelder bug. (*ent*) See LEPTOCORIS TRIVITTATUS.

Bread beetle. (*ent*) See TENEBROIDES MAURITANICUS.

Breeding potential. (*zoo*) The maximum number of offspring that an animal may produce over a specified period, or the animal's life span, given opti-mal conditions for growth, reproduction and survival of the offspring.

Bristle-tails. (*ent*) See THYSANURA.

British Pest Control Association. (*name*) The objectives of the Association (formerly the Industrial Pest Control Association) are to represent the views of the pest control industry in Britain, in safeguarding the health of the community and to conserve commodities and protect property from destructive, disease-carrying and irritating pests. To this end, the Association communicates with Government Departments and others on legislation relating to the pest control industry which may affect the interests of its members.

One of the main objects of the Association is to improve and main-tain the status of the Industry in establishing minimum standards of competence for service staff. Since its formation in 1944 the Association has ensured the safe use of its members' services and products; an obliga-tion of members of the Association is to bring to its attention information which may lead to improvement in safety standards and discourage the misuse of any practice or pesticide involving possible health hazards. To this end all members are bound by the Association's support for co-operation with Government through the PESTICIDES SAFETY PRECAUTIONS SCHEME. Detailed information about the Association is available through the Secretary, Alembic House, 93 Albert Embankment, London S.E.1.

Broad-horned flour beetle. (*ent*) See GNATHOCERUS CORNUTUS.

Bromofume. (*p. prod*) See ETHYLENE DIBROMIDE.

Bromophos. (*chem*) An organophosphorus insecticide of exceptionally low mammalian toxicity first described in 1964 and introduced in that year by Boehringer/CELA GmbH under the trade name 'Nexion'. It has stomach and contact action.

The technical product is a yellowish crystalline solid with a faint characteristic odour, soluble in most organic solvents, non-corrosive and stable under alkaline conditions.

Its full range of insecticidal activity is in process of being documented; effective against a number of insects of stored foodstuffs and especially effective against flies (0.5 g/m²).

The acute oral LD_{50} (rat) is 3.8–6.1 g/kg. Formulations available include 25% and 40% emulsion concentrates, 25% wettable powder and a thermal fogging solution (40% a.i.). The normal ready-to-use spray concentration is 0.5–1.0% active ingredient.

Brown-banded cockroach. (*ent*) See SUPELLA LONGIPALPA.

Brown cockroach. (*ent*) See PERIPLANETA BRUNNEA.

Brown dog tick. (*zoo*) See RHIPICEPHALUS SANGUINEUS.

Brown house moth. (*ent*) See HOFMANNOPHILA PSEUDOSPRETELLA.

Brown rat. (*zoo*) See RATTUS NORVEGICUS.

Brown spiders. (*zoo*) See LOXOSCELES.

Bruchids. (*ent*) See ACANTHOSCELIDES and CALLOSOBRUCHUS.

Bryobia praetiosa. (*zoo*) Acari: Tetranychidae. The Gooseberry red spider mite or Clover mite. A pest of horticultural importance which invades buildings in the spring and summer. The adult (0.5–1 mm long) is oval, either green or dull red, with slender legs, the first pair much longer than the rest. This mite has a strong migratory habit disturbing to occupants, but the mites are harmless and do not breed indoors. They are especially common in southern England and may cause alarm when they occur in considerable numbers. The mites feed on grass, clover and on fruit trees, periodically seeking the protection of buildings for moulting, and hibernation. The eggs are globular, bright red and are laid on grass. The emerging larvae feed on plant sap. *Bryobia* moves slowly, cf. BALAUSTIUM.

Preventive measures involve removal of grass within 2 feet of buildings to leave an area of exposed soil. A similar gap should be left between grass and any concrete or paving surrounding the walls. Where mites occur, spray outside walls and surrounding vegetation. See DICOFOL. Indoors, spray window frames and walls on which the mites crawl.

Bubonic plague. (*dis*) See PASTURELLA PESTIS.

Bug. (*ent*) Strictly, a member of the Order Hemiptera, e.g. the Bedbug, the Boxelder bug. A descriptive term for many small animals, e.g. 'Sow bug' (= Woodlouse, see ISOPODA), 'Pill bug' (= Millipede, see DIPLOPODA). Colloquially insects and bacteria in general.

Building demolition. (*bldg*) The destruction of buildings and site clearance prior to the use of land for redevelopment. Frequently insufficient precautions are taken by the persons responsible to prevent the dispersal of rats and mice resident in buildings being demolished, with the result that neighbouring premises, such as shops, food manufacturing premises or stores, become infested despite precautions taken by their occupants.

In the U.K. the Prevention of Damage by Pests Act (1949) gives local authorities the power to ensure that action is taken to prevent this happening: namely the extermination of rodents *before* demolition starts and the sealing off at ground level of drains, sewers and other pipes (in accordance with the Public Health Act 1961).

C

Cabinet beetles. (*ent*) See TROGODERMA.

Cacao moth. (*ent*) See EPHESTIA ELUTELLA.

Cadelle. (*ent*) See TENEBROIDES MAURITANICUS.

Cadra. (*ent*) See EPHESTIA.

Caid. (*p. prod*) See CHLOROPHACINONE.

Calandra. (*ent*) See SITOPHILUS.

Calcium cyanide. (*chem*) A fumigant introduced for insecticidal purposes as 'Cyanogas' by American Cyanamid Co. in 1923. A grey granular powder decomposed by moisture to release the gas HYDROGEN CYANIDE (*q.v.*). It has been used for the control of ground nesting wasps by introducing the powder into nest entrances and covering with soil. It performs a similar function in rodent control; the powder being introduced by long-handled spoon or GAS PUMP (*q.v.*) into rat burrows and the entrance then 'heeled-in'. Calcium cyanide together with magnesium cyanide is available for this purpose under the trade name 'CYMAG'.

Cyanide formulations should only be used by fully trained and experienced personnel who are aware of the hazards involved.

Call-back. (*name*) A colloquial term. A request by the client of a pest control contractor for his services; perhaps through failure to eradicate a pest, a repetition of the problem or some additional pest problem arising on the client's premises.

Calliphora erythrocephala. (*ent*) Diptera: Calliphoridae. Blue bottle fly, Blow fly. A pest of buildings where meat occurs; slaughter houses, canning factories, meat processors and larders of private homes. Outdoors, associated with decaying animal matter and excrement; rubbish tips.

Adults (6–12 mm long) are metallic blue, flying into dark places in search of meat for egg laying. Eggs take 24 hours to hatch; larvae are typical maggots becoming full grown in about 1 week. Pupation takes about 2 weeks, often in soil. A related fly with similar habits is *C. vomitoria*.

Callosobruchus. (*ent*) Coleoptera: Bruchidae. Cowpea beetles, Cowpea weevils. Breed only in cowpeas (*Vigna* spp.) and related legumes, infesting the crop in the field, but most damage occurs in storage. First indications of an infestation are usually adult emergence holes in the seed coat; many develop within each pea.

There are two common species, *C. maculatus* and *C. chinensis*, both similar in appearance and biology. The adults (3 mm long) are red-brown with short elytra exposing the hind third to a fifth of the abdomen. There are two dark marks on each elytron. The thorax is covered with fine white hairs. The antennae with a toothed edge.

Eggs are glued to the cowpeas. The larva feeds within the pea where it pupates. Development from egg to adult (35°C) takes 3 weeks. Development ceases below 18°C. Fumigation is the principal method of control, with attention to the cleaning and treatment of the fabric of storage.

Calorifier. (*bldg*) A heat exchange unit in which water is heated by coiled steam pipes. Common in many large buildings and often a site of infestation by cockroaches and silverfish.

Camponotus. (*ent*) Hymenoptera: Formicidae. Carpenter ants. A genus of large, conspicuous ants (workers 6–10 mm long), brown-black, the pedicel with one knob. Pests of warm climates; absent in Britain except when occasionally introduced. Galleries are excavated in sound and unsound wood for nesting; often producing extensive structural damage to properties.

The galleries are distinguished from those of termites by the absence of debris, faecal pellets and the clean surface to worked timber. Properties near woodland are especially vulnerable to invasion; damp timbers are most readily attacked. Control measures include the treatment of galleries with insecticidal dusts and irrigation with oil-based solutions. Outdoors, nests in the vicinity of infested properties should be destroyed; underfloor areas and the soil within a few feet of buildings should be sprayed.

Carabidae. (*ent*) The family of Ground beetles; casual visitors to warehouses, granaries and sometimes domestic properties, rather than true residents. They are usually found in ones or twos, rarely in larger numbers, in dark corners, under linoleum, in cellars and other little disturbed areas. Small numbers (occasionally hundreds) of adults may enter houses but their invasions are restricted to a few months of the year.

Outdoors they live under stones, timber or wherever there is shelter. When touched they run quickly to the nearest hiding place. The majority are beneficial as predators on many injurious insects and slugs.

The adults are flat, the larger species being up to 25 mm long. Colour is very variable: black, brown, or metallic green, blue or violet. Antennae long and filiform.

Treatment is not normally necessary; the occasional intruder may be picked up and destroyed. Otherwise use insecticidal dusts where the adults collect and treat points of entry and wall-floor junctions.

Carbamate insecticides. (*chem*) A recently developed group of insecticides. They act as inhibitors of the enzyme cholinesterase, but in contrast with organophosphorus insecticides, the inhibition of cholinesterase is reversible. This allows rapid recovery from poisoning in man and also allows the recovery of insects from sub-lethal doses. Examples of the group are PROPOXUR ('Baygon') and CARBARYL. The Carbamates are not readily soluble in the organic solvents generally used for formulating oil-based insecticides. They are readily absorbed into the body by all routes and have a wide spectrum of insecticidal activity.

Carbaryl. (*chem*) An insecticidal carbamate, its properties first described in 1957 and introduced by Union Carbide Corp. under the trade name 'Sevin'. A white crystalline solid, stable to light and heat, and non-corrosive. A useful contact insecticide of low toxicity to man, particularly effective against Hymenopterous pests (ants, wasps) and for safe use in and around the home against casual intruders. Also effective against parasites of domestic animals (fleas and ticks) and for the control of poultry mite. The acute oral LD_{50} (rat) is 850 mg/kg. The usual formulations are 50% wettable powder (applied as 5% spray) and dusts containing 5–10% active ingredient.

Carbofos. (*p. prod*) See MALATHION.

Carbohydrate. (*chem*) A group of chemical compounds, composed of carbon, hydrogen and oxygen, containing simple and complex sugars, hemicelluloses and starch.

Carbon disulphide. (*chem*) Carbon bisulphide. An inorganic fumigant first used as an insecticide in 1854. It is a colourless to yellowish liquid containing impurities with an unpleasant odour. The vapour is extremely inflammable (flash point 20°C) and ignites spontaneously at 130°C. The density of the vapour is 2.63 to that of air.

It is used as a fumigant for nursery stock and for the treatment of soil against insects and nematodes. This fumigant is also used in mixtures with CARBON TETRACHLORIDE (*q.v.*), to reduce fire risk, for the treatment of stored grain. It has high penetrative properties.

The vapour is highly poisonous causing giddiness and vomiting in 30 minutes at a concentration of 1:500. Ill health is caused by repeated daily exposure to concentrations of 1:15,000. High concentrations produce narcosis, when absorption may occur through the skin.

Carbon tetrachloride. (*chem*) 'Carbon-tet', CTC. An insecticidal fumigant, first used for the treatment of nursery stock in 1908; a colourless, sweet-smelling, non-flammable liquid, about 5 times heavier than air, vaporising easily at room temperature.

Its insecticidal properties are low, being used for the treatment of grain when long exposures are possible; absorption into treated grain is small. It fills a useful role as an ingredient of fumigant mixtures, principally for grain, with e.g. ETHYLENE DICHLORIDE and CARBON

DISULPHIDE (*q.v.*) to reduce their fire risk. Carbon tetrachloride is a general anaesthetic; prolonged exposure causes irritation of mucous membranes, headache and nausea and extensive damage to the liver. The acute oral LD_{50} (rat) is 5730–9770 mg/kg, but in practice, poisoning of man most often occurs from repeated inhalation. The threshold limit is only 10 ppm for continuous daily exposure.

Carpenter ants. (*ent*) See CAMPONOTUS.

Carpet beetles. (*ent*) See ANTHRENUS VERBASCI and ATTAGENUS.

Carpophilus. (*ent*) Coleoptera: Nitidulidae.
A genus of Dried fruit beetles. Serious pests of currants and raisins, and often cereals, nuts and spices. Pests in warehouses, and outdoors on ripe and decomposing fruits, the larvae spoiling the crop with frass and cast skins. Flat, oval beetles (2–4 mm long) contaminants of products imported into Britain. Examples are *C. ligneus* and *C. dimidiatus*, with elytra uniformly brown: common in central and north America. Also *C. hemipterus* (right), elytra brown, with a patch of yellow on the shoulders and apex: especially troublesome in California. Development to adult about 3 weeks (28°C).

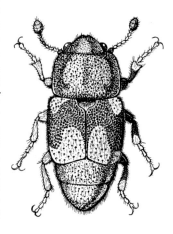

Carrier. (*chem*) See DILUENT.

Case-bearing clothes moth. (*ent*) See TINEA PELLIONELLA.

Castes. (*ent*) Members of a species performing different functions within a colony and often strikingly different in appearance. Pests which exhibit a well-developed caste system are ants and wasps (Hymenoptera) and termites (Isoptera): e.g. in termites the workers (sterile females) performing nest building, foraging and feeding of young (Fig. 16); soldiers protecting the nest from predators (Fig. 14); alates (sexual males and females), contributing to dispersion and the establishment of new colonies (Fig. 15), and queens (fertilised females) producing young. Termites are unique among insects in the extent to which the production of castes is displayed, probably accounting for the large size to which some species are able to develop colonial populations.

The existence of foraging workers among the ants and wasps is exploited in the use of insecticidal baits which are carried back to the nest to kill the queen and developing young.

Castor bean tick. (*zoo*) See IXODES RICINUS.

Castrix. (*p. prod*) See CRIMIDINE.

Casual intruders. (*zoo*) Insects and other animals which enter buildings fortuitously, sometimes in their search for harbourage, or a place to hibernate, or at night by attraction to light. They normally fail to reproduce within buildings, occurring as individuals, or at most in twos or threes. Casual intruders may cause annoyance or distress, but do not damage the building or its contents and are not carriers of disease.

Catering premises. (*bldg*) Buildings (hotels, motels, restaurants, cafés) in which food is provided and of special significance in pest control. They 1) are subject to inspection by public health departments, 2) handle a wide variety of foods subject to infestation, 3) obtain materials from many suppliers also subject to infestation, 4) provide food for direct public consumption, 5) have a reputation to maintain for the quality and wholesomeness of the meals provided, 6) are situated very often in areas of high population density and therefore high rodent risk, 7) are usually equipped with bars often conducive to cockroach infestation.

Treatment for pest control is often complicated by the hours of business of some catering premises, viz. 24 hours a day.

Caterpillar. (*ent*) A non-technical term: strictly the larva of a butterfly or moth (Lepidoptera), although commonly also used for the immature stage of beetles and other insects in which the larva has obvious legs and is actively mobile. See GRUB and MAGGOT.

Cat flea. (*ent*) See CTENOCEPHALIDES FELIS.

Cavity wall. (*bldg*) A wall consisting of two thicknesses of brickwork, or other building material, with a cavity between (usually about 5 cm wide), the object being 1) to prevent the transfer of moisture into a building and 2) the air space to provide insulation against heat loss. Through faulty construction, broken air bricks or badly sealed piping, access to these cavities is often provided for insect and rodent pests. The liberal use of insecticidal and rodenticidal contact dusts provides the easiest means of control, together with effective sealing of all openings into the cavity.

Cellar beetle. (*ent*) See BLAPS MUCRONATA.

Census baiting. (*proc.*) See TEST BAITING.

Centipedes. (*zoo*) See CHILOPODA.

Ceratophyllus. (*ent*) Siphonaptera: Cerato-phyllidae. Bird fleas. Common parasites of wild birds infesting nests in dry locations, occasionally attacking man. *C. gallinae* is the most common bird flea in the U.K.; recorded from about 75 different hosts, including sparrows, starlings and poultry.

Chafers. (*ent*) Coleoptera: Scarabaeidae. Plant feeding beetles, which occasionally enter buildings, e.g. the Cockchafer, or May bug, *Melolontha melolontha* (right). This is a minor pest of agriculture in the U.K., but a major pest of grassland on the Continent. The adults emerge in May, fly well, and are attracted to artificial light at night, often blundering into obstacles.

Harmless: do not bite or sting despite their 'impressive' armoured appearance. The adult (20–25 mm long) is a heavy-bodied insect, with blac, prothorax clothed with white hairs, yellow-brown elytra, the end of the abdomen bent downwards. The antennae composed of many tiny plates (lamellae). Control methods are not usually necessary; proofing is effective against this and other night-flying insects.

Cheese mite. (*zoo*) See TYROPHAGUS CASEI.

Chemosterilant. (*chem*) A chemical capable of causing sexual sterility in insects, birds or mammals by preventing or suppressing reproduction, but not influencing the size of the population at the time it is used. Such materials are likely to exert less effect on other biological organisms in the environment compared with pest-icides which kill.

Chilopoda. (*zoo*) The Class of animals re-ferred to as centipedes, with elongate somewhat flattened bodies. Most seg-ments are similar, each with one pair of legs (cf. DIPLOPODA). The number of segments varies with the species from 15–90. A pair of poison claws on the first body segment is used to inject venom into prey; in Britain the venom is unlikely to harm human beings, but in some tropical countries, attack by centipedes can be very painful.

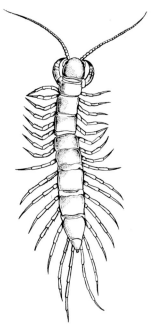

Centipedes are always found in damp locations, but they may crawl indoors for shelter, under doors, through gaps in windows or up water pipes into kitchens and bathrooms.

Common species in Britain are *Necro-phlaeophagus longicornis* (40 mm), yellow, its body usually looped and winding in motion, and *Lithobius forficatus* (30 mm) dark brown, shiny and with a stouter body. Eggs are laid in batches and are tended by the female until hatching. Treat-ment of buildings is rarely necessary.

Chiroptera. (*zoo*) Bats. A specialised group of mammals adapted for flying by a membrane extending from the front legs along the sides of the body to the hind legs and tail. By day they roost in dark places (roof voids, belfries, hollow walls, in cavities in trees) hanging upside down, often congregating in large colonies, and fouling structures beneath. Their significance as pests arises from the odour of bat dung, the insect infes-tation it encourages, noise created in rooms below and the staining of ceilings. Two groups are recognised: the insect-eating and fruit-eating bats. Not pests of any significance in the U.K.; they reach their greatest numbers in tropical climates; rabies is recognised in insectivorous bats in some countries. FUNGAL INFECTIONS (*q.v.*) may also arise from infected guano.

Recognised methods of control involve the proofing of buildings, use of repellents and toxic dusts.

Chloralose. (*chem*) See ALPHACHLORALOSE.

Chlordane. (*chem*) An organochlorine insecticide discovered in 1945 and introduced in that year by Velsicol Chemical Corp. under the trade name 'Octachlor'. The technical product is a viscous amber liquid, readily miscible with most organic solvents, consisting of 60–75% isomers of chlordane and 25–40% of related compounds including two isomers of HEPTACHLOR (*q.v.*).

This insecticide has persistent stomach and contact activity, drying to give a resinous film which is effective against a wide range of insects. As a result, chlordane has become one of the most widely used insecticides for industrial and domestic pest control especially for cockroaches, ants and other household pests; in mists and fogs against mosquitoes; and for soil poisoning against termites. See TERMITE CONTROL.

Extensive use of chlordane resulted in resistance to this insecticide being detected in German cockroaches in 1953 (see CORPUS CHRISTI STRAIN) and now prevalent in many North American States, with CROSS-RESISTANCE (q.v.) to a number of other organochlorine insecticides. Brown dog ticks (*Rhipicephalus sanguineus*) may also be resistant in some areas.

Chlordane is readily absorbed through the skin; it is somewhat volatile affording limited vapour action, with a sickly characteristic odour. It should not be used extensively in ill-ventilated areas. The acute oral LD_{50} (rat) is 457–590 mg/kg. Formulations most commonly available include oil concentrates, ($\simeq 20\%$), emulsion concentrates (50–70%), wettable powder, and dust (5–10%).

Chlordecone. (*chem*) A chlorinated ketone, introduced as an insecticide in 1958 by Allied Chemical Corp. under the trade name 'Kepone'. It is chemically unlike other chlorinated hydrocarbon insecticides; the technical form is an off-white crystalline solid, odourless and highly stable, with low volatility.

The major use of chlordecone is in baits (0·125–0·25% active ingredient) for the control of cockroaches, ants and crickets. Baits provide a long-lasting, but slow acting stomach poison requiring 4–10 days to take effect. They are extremely useful in areas (e.g. animal laboratories, zoos (Fig. 18), pet shops, aquaria) where insecticidal dusts and sprays cause hazard. Baits are formulated as pelleted cereals, pastes and gels.

The acute oral LD_{50} (rat) is 125 mg/kg; safety in use derives from the incorporation of chlordecone at low concentrations and placement in locations favoured by insects but not readily accessible to man.

Chlorinated hydrocarbons. (*chem*) A group of numerous related compounds characterised by their containing chlorine, hydrogen and carbon, and a stability which gives them a long residual life. They do, however, differ widely in chemical structure, activity and mammalian toxicity from, for example, endrin LD_{50} (rat) 11 mg/kg to methoxychlor 6000 mg/kg.

They are more dangerous to humans as acute rather than chronic poisons. Many are stored in body fat where they appear largely inactive, since the body can store more than what would be a lethal dose if given at one time. Most are absorbed through the skin, chlordane and dieldrin especially rapidly.

It is this group of insecticides which has been criticised for long persistence in the environment and for this reason, use is being made increasingly of organophosphorus and carbamate insecticides as replacements.

Chlorophacinone. (*chem*) An indane-dione type anticoagulant rodenticide, introduced in 1961 by Lipha SA, under the trade names 'Caid', 'Liphadione', and 'Raviac'; also as 'Drat' (May and Baker), 'Quick' (Rhone-Poulenc) and 'Rozol'.

It is an odourless, slightly yellow crystalline solid, soluble in organic solvents but only sparingly soluble in water. Non-corrosive. Chlorophacinone is incorporated into rodenticidal baits at 0·005–0·025%; a single dose (0·005%) is said to kill brown rats from the fifth day after feeding. LD_{50} (rat) at 6 days, 2 mg/kg; wild *Rattus norvegicus* at 6 days, 5 mg/kg. Acute oral LD_{50}, 21 mg/kg.

Chlorophacinone is claimed to have a smaller anticoagulant effect on human blood than warfarin, and is less toxic to cats, dogs, and pigs.

Available as a concentrate in oil solution (0·25%) and as prepared baits (0·005%).

Chloropicrin. (*chem*) Nitrochloroform. A non-flammable, colourless liquid, miscible with most organic solvents, and corrosive to iron, zinc and other light metals. Its use as an insecticide was patented in 1908, finding application as a fumigant for the treatment of stored grain, and against insects, nematodes and some fungi in soil. Mixed 50% with motor oil, chloropicrin is used as a fumigant in rat burrows with continuing repellent action.

Its main use has been as a warning gas at low concentration with other fumigants (e.g. methyl bromide) when it has active lachrymatory properties (tear gas), and causes intense irritation of the mucous membranes. Used direct it is generally applied at 30–80 g/m^3. Chloropicrin is lethal to man after a 30-minute exposure to 0·8 mg/litre of air. This highly toxic property is offset by the obvious irritant effects evident at much lower concentrations (less than 1 ppm in air).

Chlorpyrifos. (*chem*) An organophosphorus compound recently introduced under the trade name 'Dursban' by Dow Chemical Co., with good insecticidal properties against cockroaches, but of slow knockdown, no flushing action and limited persistence. Also for mosquito control.

The technical material is a white crystalline solid of low volatility with a mild sulphide odour; insoluble in water, but soluble in most organic solvents. Rapidly degraded by alkaline surfaces. Compatible with knockdown agents such as pyrethrins and dichlorvos.

The acute oral LD_{50} (rat) is 145 mg/kg; dermal hazard is low. Formulations available include emulsion concentrates, recommended for use against cockroaches at 0·25–0·5%.

Choice test. (*proc*) See TEST METHODS.

Cholinesterase. (*chem*) Acetylcholinesterase, AChE. An enzyme of the body essential for its normal functioning. In cases of poisoning, organophosphorus insecticides act as irreversible inhibitors, combining with and inactivating the enzyme allowing the accumulation of large amounts of acetylcholine. The severity of poisoning derives from 1) the degree of cholinesterase depression and 2) rate of depression. Cholinesterase values are determined on samples of blood by special techniques, measurements being taken on the red cells and plasma.

Recovery becomes complete if a victim has time to regain his cholinesterase level: this is accelerated by the use of antidotes which break the bond between the organophosphorus compound and cholinesterase, e.g. PRALIDOXIME CHLORIDE (*q.v.*).

Chromatography. (*proc*) See ANALYTICAL METHODS.

Chronic poison. (*tox*) A pesticide which is used at low concentrations and relies on repeated intake over a period of time to achieve its effect, e.g. most anticoagulant rodenticides. Cf. ACUTE POISON.

Chrysopa carnea. (*ent*) Neuroptera. The Green lacewing. An innocuous intruder of homes where it hibernates in winter. It is attracted to light at night in the autumn; beneficial as a larval predator on aphids outdoors. The adult is characterised by a long slim body; the green-yellow wings, extensively veined, fold tent-wise when at rest. It has large shining, brightly coloured eyes and long slender antennae (see right).

Churchyard beetle. (*ent*) See BLAPS MUCRONATA.

Cigarette beetle. (*ent*) See LASIODERMA SERRICORNE.

Cimex lectularius. (*ent*) Hemiptera: Cimicidae. The Bedbug. At one time a common pest of slum and other poor properties, but now much reduced by higher standards of hygiene. Still common, however, in apartment buildings, hotels and motels, with a rapid turnover of residents. Always closely associated with man who is its only

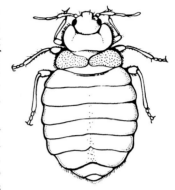

host. Regarded by most with abhorrence, not a carrier of disease. Bedbugs feed at night and are not evident during the day, hiding in cracks and crevices of walls, mattresses, bed springs and head boards. The eggs are cemented to surfaces, close to a host. One blood meal is taken during each nymphal stage.

The adult (5 mm long) is oval, flattened, red-brown, without wings and can live several months without food. Search for a host may be in response to warmth. A complete life cycle takes 2–4 months. Stink glands produce an objectionable smell. Control by fumigation is now largely replaced by use of residual insecticides.

C. hemipterus, the Tropical bedbug occurs with *C. lectularius* in warmer climates.

Circadian rhythm. (*ent*) A built-in cycle of activity established by heredity; a pattern of behaviour of many insects with a peak of activity at a certain time of the day repeating itself 24-hourly, although environmental conditions may change (*circa* = about, *diem* = day).

Well-illustrated by pest cockroaches: a marked increase in activity occurs with onset of darkness, which is maintained for about 5–6 hours, the insects then remaining quiescent in harbourages throughout the following day.

Circulatory system. (*zoo*) Of insects, the dorsal blood vessel and the haemocoel (body cavity) containing the haemolymph and various blood cells (haemocytes) with no oxygen-carrying function; the medium for the transport of digestive products, regulatory substances (hormones) and waste salts. In mammals and birds, the heart, arteries, veins and finer capillaries (a closed system) carrying blood, with oxygen carrying functions, digestive products to, and waste materials from, the body tissues.

Classification. (*zoo*) A system of convenience devised by biologists; an ever changing arrangement whereby animals (and plants) with the closest similarity of form, physiology and behaviour are *grouped* to provide 'order' among the wide range of forms that exist, and *divided* into smaller units to make easy the naming of species in any group. An artificial concept, which establishes relationships between organisms, and provides an aid to devising keys for identifying species.

Cleaning behaviour. (*zoo*) See GROOMING.

Clean out. (*name*) A colloquial term, used commonly in the United States, for a single treatment, usually for cockroach control, often associated with a 'guarantee' that free service will be provided if the problem recurs in a specified time. See PEST CONTROL CONTRACT.

Clothes moths. (*ent*) See TINEA and TINEOLA.

Clover mite. (*zoo*) See BRYOBIA PRAETIOSA.

Cluster flies. (*ent*) Diptera. Swarming house flies, which hibernate in large numbers, entering homes and roof spaces especially in September–October, causing a nuisance to occupiers but leaving again in the Spring. Mainly members of the Muscidae, especially *Pollenia rudis*, *Musca autumnalis* and also *Thaumatomyia notata* (Chloropidae). Smoke generators are useful in the treatment of roof voids; sprays should be used around window frames.

Coaming. (*ship*) A raised structure around the edge of the hatch of a ship or barge to keep the water out. On 'open' lighters (unhatched) the coaming supports the edges of fumigation sheets weighed down to the deck with chains to provide an effective seal. SAND SNAKES (*q.v.*) are not used.

Coccinellidae. (*ent*) The family containing the well-known Ladybird beetles; oval, strongly convex beetles with conspicuous markings of red, black, yellow and white. Casual intruders of homes where they sometimes hibernate; rarely a pest except when present in large numbers. Common British species are *Coccinella septempunctata* (red elytra with seven black spots) and *Adalia bipunctata* (of variable colour, often red with two black spots).

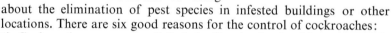

Cockchafer. (*ent*) See CHAFERS.

Cockroach control. (*proc*) The application of insecticidal formulations to bring about the elimination of pest species in infested buildings or other locations. There are six good reasons for the control of cockroaches:

1) Cockroaches, or parts of their bodies, in food sold to the public result in considerable loss of business and goodwill. Fines may also be imposed following prosecution. The resulting publicity is particularly damaging to reputable companies.

2) Food fouled and tainted by the characteristic odour of cockroaches is unfit for human consumption. Odours may remain even after cooking or processing.

3) Cockroaches can carry a number of diseases injurious to man and of importance in public health, notably *Salmonella* infection.

4) The presence of cockroaches causes unnecessary distress and psychological harm to many members of the public.

5) Cockroaches are omnivorous; they damage many articles other than foods, i.e. books, tapestries, leather goods, pictures.

6) Relatively few people will tolerate cockroaches at their place of work. Some may even assert that cockroaches have attacked them; working conditions are an important consideration in the retention of employees.

Cockroaches. (*ent*) See BLATTARIA.

Cocoa moth. (*ent*) See EPHESTIA ELUTELLA.

Coleoptera. (*ent*) Beetles. The largest Order of insects. Adults have the forewings modified and hardened to form wing cases (elytra), which are sometimes fused and usually cover the abdomen; membranous hindwings are folded beneath, sometimes absent. The body is often heavily chitinised; the mouthparts usually modified for biting, the prothorax well-developed and mobile. Variation in species size is considerable. The larvae usually with legs (except for the weevils). Pupal stage present.

Many families contain pest species. Of considerable economic importance are the Dermestidae, Anobiidae, Ptinidae, Curculionidae, Nitidulidae, and Tenebrionidae. Between them, they infest stored foodstuffs, textiles and timber. Larvae of many beetles are scavengers or predators. Adults of some are often casual intruders in the home.

Collembola. (*ent*) Springtails. Small primitive wingless insects, prevalent in soil, scavengers on organic matter, occasionally getting indoors; into damp kitchens, cellars and outbuildings, around sinks and drains. Characterised by a paired leaping organ on the end of the abdomen. Springtails do no damage, but are sometimes a nuisance if present indoors in large numbers.

Colony. (*zoo*) A number of organisms of one kind living together as a group, e.g. of cockroaches in a harbourage, or of rats on a refuse tip.

Columba livia. (*zoo*) Pigeon, Feral pigeon. A descendant of the rock dove. The most important pest bird of cities, moving in large flocks, roosting on the ledges of buildings at night, defacing and eroding the stonework by their droppings, causing unsightly areas of fouling on pavements, in the entrance to buildings, station platforms and other public areas (Figs. 17 and 21). Common pests also around grain terminals on port docksides where they feed on spillage.

To some extent pigeons in cities have been encouraged in recent years by the public feeding them in parks and recreational areas: this is now an offence in some countries. The principal food is seed and green feed, but in cities they scavenge for whatever food is available, or provided, close to roosting sites.

Nests are constructed on buildings, in recesses, statuary, sheltered ledges, behind parapets; there are usually 2 eggs per clutch, 2–3 broods each year, hatching time 2–3 weeks, the fledglings leaving the nest after a further 5 weeks.

Control methods in the U.K. consist of proofing, use of tactile repellents and narcotising under licence to the M.A.F.F. See BIRD LAWS, BIRD CONTROL, ORNITROL and AVITROL.

Commercial. (*name*) A term used by pest control contractors to distinguish a category of clients, or source of business, (incorporating industrial) as distinct from domestic.

Commodity. (*manuf*) Goods, produce. A raw material or finished product handled in trade, often carrying infestation with it from one location to another, varying in susceptibility to pest attack with climate and the quality of storage.

Common clothes moth. (*ent*) See TINEOLA BISSELLIELLA.

Common furniture beetle. (*ent*) See ANOBIUM PUNCTATUM.

Common rat. (*zoo*) See RATTUS NORVEGICUS.

Common sheep tick. (*zoo*) See IXODES RICINUS.

Concentrate. (*chem*) A formulation containing a pesticide at a higher level than is normally used, but in a form suitable for dilution. It incorporates solvents and emulsifiers (for sprays) and diluents (for dusts and baits). For example a concentrate of warfarin (1%) for admixture with a food base to give a 0·005–0·05% ready-to-use bait; an emulsion concentrate of lindane (20%) for dilution with water to give a ready-to-use spray of 1%.

Concentrates are formulations of 'convenience'; they reduce costs and the bulk to be carried; because they contain a high level of active ingredient they should always be stored in a safe place and handled with care, especially in transport and during dilution.

Confused flour beetle. (*ent*) See TRIBOLIUM CONFUSUM.

Contact dust. (*chem*) A rodenticidal dust, used to treat dry harbourages and runs (Fig. 19), picked up on the feet and underside of the body during movement, passing directly through the skin, or into the mouth during grooming. Often wrongly called TRACKING POWDER in some countries. Examples of chemicals used in this way are warfarin (1%), DDT (20–50%) (now banned from use in many countries), and lindane (50%). Because of relatively high toxicity, contact dusts must not contaminate exposed surfaces.

Contact insecticide. (*chem*) An archaic term which arose in agricultural applications of insecticides to distinguish those compounds which killed insects by coming into contact with treated leaves and those which were effective against leaf-eating species.

In industrial/domestic use, a chemical which kills when the pest runs over or alights on a treated surface. The insecticide may act either directly by penetration of the body wall or by ingestion. See STOMACH POISON.

Contact test. (*proc*) See TEST METHODS.

Container. (*name*) A structure in which goods are enclosed for transport. Of shipping, a very large box holding many tons of goods, introduced (as 'containerisation') for speeding up the loading and unloading of cargoes at ports. Such containers are used repeatedly for different consignments, may have lined walls providing cavities for pest harbourage and are usually sufficiently air-tight for the treatment of contents by fumigation (e.g. Phostoxin).

Contamination. (*chem*) Pollution, defiling by admixture; for example, the inadvertent spraying of foodstuffs with an insecticide (Fig. 20), the thoughtless disposal of a pesticide container with residues in a stream; the treatment of a rubbish tip with insecticidal dust or spray on a windy day, so that drift occurs onto adjacent land.

A subject of special importance to all involved in pest control, 1) as representatives of a responsible industry, 2) conforming with legislative requirements for the safe use of pesticides, 3) recognising the public interest in pesticides, regarding the wholesomeness of foods and the safeguarding of the environment. (See also Fig. 29.)

In food manufacture, the contamination of products by insects or parts of their bodies, rodent droppings and rodent hairs, resulting in loss of GOODWILL (*q.v.*).

Control of infestation. (*proc*) See ERADICATION.

Conveyor. (*manuf*) Equipment for the transfer of materials and finished goods, usually horizontally (see also ELEVATOR). Four basic types are common, presenting different infestation problems:

1) Belt conveyor: an endless belt supported on rollers often carrying raw commodities (e.g. grain) at speeds up to 400 tons/hr; residues and spillage are minimal; the commodities are often treated with insecticidal sprays and dusts (e.g. malathion) from nozzles or jets mounted over the belt.

2) Screw conveyor: a spiral 'worm' on a central axis enclosed in a channel, moving the commodity along by rotation; spillage is minimal but residues in the channel are often high. Frequent cleaning is required if infestations and webbing (e.g. of *Ephestia kühniella*) are not to accumulate in the 'dead space'. SPOT TREATMENT with pyrethrins may be required.

3) Drag chain conveyor: a chain with metal structures attached and enclosed in a channel, the commodity dragged along its length; spillage is minimal, but residues in the channel are often high. Pest problems as in 2) above.

4) Roller conveyor: a series of rollers put into motion by goods (usually packaged commodities) being given impact at one end, or fed onto the conveyor by gravity. Spillage and residues are minimal. These conveyors are often supported on hollow metal stanchions and are frequently used in wet areas (e.g. dairy product manufacture and bottling works). They provide excellent harbourage for cockroaches.

All conveyors powered by electric motors provide local sources of heat; the casings often provide insect harbourages which are inaccessible to treatment.

Cooling tunnel. (*manuf*) A long closed structure, common in many food manufacturing industries, through which cooked products (e.g. bakery and confectionery) are passed for controlled or rapid cooling. Often containing food residues, frequently supporting infestations of *Tribolium* and mice, and without access for treatment.

Copra beetle. (*ent*) See NECROBIA RUFIPES.

Corn thrips. (*ent*) See LIMOTHRIPS CEREALIUM.

Corpus Christi strain. (*ent*) A strain of German cockroaches resistant to chlordane bred from a population of insects taken from Corpus Christi,

Texas in 1953. In late 1951 and early 1952 reports from Texas indicated failures to control the German cockroach satisfactorily with chlordane. High level resistance to chlordane in a natural population was later confirmed; also that resistance extended to other chlorinated hydrocarbon insecticides of the CYCLODIENE type (*q.v.*). Resistance by the German cockroach to these insecticides is now widespread in the U.S.A., and on ships. It has also been recorded among cockroaches in Europe.

The Corpus Christi strain is maintained in many laboratories of the world as a 'standard' against which new insecticides are evaluated.

Co-solvent. (*chem*) A liquid used to increase the solubility characteristics of a solvent used as the 'carrier' for a pesticide. See SOLVENT and SOLUBILITY.

Coumachlor. (*chem*) An anticoagulant rodenticide of the hydroxy-coumarin type introduced in 1953 by Geigy under the trade marks 'Tomorin' and 'Ratilan'. The technical product is a stable, yellow crystalline powder (about 70% pure) practically insoluble in water, slightly soluble in some organic solvents. Formulations available include a 'tracking powder' (1%), bait concentrates (0·5% and 1%) and as Ratilan bait blocks.

Coumafuryl. (*chem*) See FUMARIN.

Coumatetralyl. (*chem*) An anticoagulant rodenticide of the hydroxy-coumarin type, introduced in 1956 by Bayer under the trade name 'Racumin'. The technical material is a yellow-white crystalline powder, tasteless and practically odourless, insoluble in water and not readily soluble in organic solvents. Formulations available include a 'tracking powder' (0·75%), which may also be diluted (1:19) for ready-to-use baits (0·0375%).

Cowpea beetle/weevil. (*ent*) See CALLOSOBRUCHUS.

Crash barrier. (*bldg*) Within buildings, a strip of metal attached to a wall to prevent damage to the surface behind. Damaged or poorly constructed surfaces in food manufacturing premises result in accumulations of food to support infestations of insects and mice. Such surfaces cannot be effectively cleaned. Less damage and therefore easier cleaning can be achieved with the sensible use of crash barriers, and other protective devices. Crash barriers also prevent the stacking of goods in contact with walls facilitating inspection for pests. See STORAGE.

Crataerina pallida. (*ent*) Diptera: Hippoboscidae. Martin or Swallow fly. A parasitic fly inhabiting birds' nests (Sand martin, House martin, Swallow and Swift), the adult feeding on blood and occasionally attacking man indoors. Not of typical fly appearance; (6 mm long) pale green, with long legs and very short wings; runs actively, not capable of flight. Viviparous; each female depositing a single fully-developed larva which immediately pupates. The pupa remains in the nest of the host bird to await its return next season.

Crickets. (*ent*) Orthoptera: Gryllidae and Gryllotalpidae. Nuisance pests, the chirping of the males, produced by the front wings being rubbed together, causing annoyance to householders.

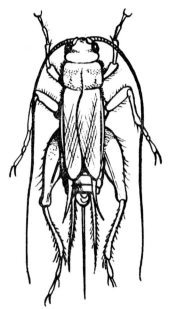

Characterised by the hind legs being well-developed for jumping; the membranous wings covering and often extending beyond the abdomen as two 'spikes'; the female with a long ovipositor.

The house cricket (*Acheta domesticus*) breeds indoors and out. Various field crickets live solely outdoors but in warm climates may occasionally enter homes. Mole crickets (*Gryllotalpa*) may damage crops and garden plants and also enter properties as casual intruders.

The house cricket (*A. domesticus*), 16 mm long, yellow-brown with darker markings on head and thorax, lives in buildings throughout the year, favouring warm locations, behind old types of fireplace, bakery ovens and heating plant. It hides in crevices by day and is active at night. Eggs laid in crevices take 2–3 months to hatch; development to adult requires 7–8 months. Properties situated close to rubbish tips are occasionally invaded by large numbers of crickets during the summer. They are omnivorous, but damage to household items (paper, food and linens) is usually minor.

Field crickets are occasional invaders of homes in warm climates. The field cricket of the United States (*Acheta assimilis*) is 20 mm long, dark brown to black and may invade buildings towards the end of the summer, but is unable to survive indoors through the winter.

Mole crickets occur in tropical and sub-tropical latitudes (e.g. Southern United States): they spend most of their time in burrows, are active above ground at night and may enter buildings via basements. The front legs are short and modified for burrowing.

Control measures for field and mole crickets are rarely necessary indoors. House crickets may be controlled with sprays, dusts and baits as used against cockroaches. Rubbish tips should be treated where infestations cause annoyance taking care not to contaminate gardens of adjacent properties.

Crimidine. (*chem*) A pyrimidine compound introduced in the early 1940's by Bayer as an acute rodenticide under the trade name 'Castrix'. The technical product is a brownish wax practically insoluble in water, but soluble in organic solvents.

Highly toxic; rapidly producing convulsions. Acute oral LD_{50} (rat) is 1·25 mg/kg, quickly metabolised not producing secondary poisoning. Product available: Castrix Giftkorner 0·1% impregnated grains, used for the control of mice and voles.

Cross-infestation. (*manuf, ship*) The movement of pests from one commodity to another by infested and clean goods being transported or stored together, or goods susceptible to infestation being carried or held in transport or warehouses the structure of which is infested.

Cross-resistance. (*zoo, chem*) The ability of a pest which has developed RESISTANCE (*q.v.*) to one pesticide, to withstand also the effects of another related chemical; e.g. cross-resistance between chlordane and dieldrin among *Blattella germanica* in the U.S.A.; also between warfarin and related anticoagulant compounds among *Rattus norvegicus* in the U.K.

Cryptococcosis neoformans. (*dis*) See FUNGAL INFECTIONS.

Cryptolestes. (*ent*) Coleoptera: Cucujidae. Flat grain beetles. Many species are minor pests of grain and cereal products, in warehouses, mills and their machinery. They do not attack sound grain but are able to enter grains which are damaged.

Tiny (2 mm long) flattened bodies, light red to brown with long antennae and a curious swaying gait. The most common species in the U.K. is *C. ferrugineus* (Rust-red grain beetle).

Eggs are laid in crevices or on food material. Minimum period of development from egg to adult is 5 weeks (27°C).

Cryptotermes brevis. (*ent*) See DRYWOOD TERMITES.

CTC. (*chem*) See CARBON TETRACHLORIDE.

Ctenocephalides. (*ent*) Siphonaptera: Pulicidae. *C. felis* (Cat flea); *C. canis* (Dog flea). Similar in appearance, sometimes attacking man, usually around the ankles, occasionally found on rats. Both are intermediate hosts of the dog tapeworm (*Dipylidium caninum*) which may infect man. Eggs are laid a few at a time on a host but soon drop off. The larvae 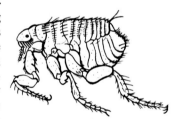 are active for 2–4 weeks, feeding on faeces of the adult and other debris; pupation occurs in cracks and crevices of the home and the adult can live several weeks without food. Breeding sites are usually concentrated around sleeping areas of pets, the basket, bedding or kennel, but may also occur in furniture.

CT product. (*chem*) Concentration × time product. The amount of a fumigant acting on insects over a certain period of time (mg hr/l). For example, the required CT product for 99% kill of *Tenebroides mauritanicus* is 166 mg hr/l. This may be obtained by a 2-hour exposure of 83 mg/l, 4 hours of 41·5 mg/l or 10 hours of 16·6 mg/l. In practice, however, fumigation dosages are usually quoted in ounces/1000 ft³. for a stated period, usually 24 hours. The dosage reduces with increasing temperature because of the higher metabolic and respiratory activity of insects.

Culex. (*ent*) Diptera: Culicidae. A large genus of mosquitoes, widely distributed, not carriers of malaria but able to transmit diseases such as encephalitis. Eggs are laid in clusters or 'rafts' without air floats (cf. ANOPHELES). Larvae hang at an angle from the surface of water; the pupae have cylindrical respiratory trumpets. Adults rest with the abdomen parallel to the surface. Some of the most important species are:

Culex pipiens: the common mosquito of towns and rural areas, frequently entering houses to hibernate. Common in the U.K. and a carrier of St. Louis encephalitis in the United States. Eggs are laid on small areas of entrapped water as well as open sites. Breeding is continuous during warm weather; development to adult is completed in 8 days. A related species in Britain which bites man is *C. molestus.*

Other species of importance in the U.S. are:

C. salinarius: common along the Atlantic coast breeding in fresh and brackish water. Adults fly up to 8 miles from breeding sites.

C. restuans: breeds in water fouled with decaying vegetation (pools, ditches, as well as tin cans). Adults disperse up to 3 miles.

C. tarsalis: an important vector of St. Louis encephalitis and western equine encephalitis, transmitted from birds to horses and man. This mosquito is widely distributed west of the Mississippi River. Eggs are laid in a variety of water situations and breeding continues throughout warm weather. Adults are usually found within a mile of breeding sites.

Culicidae. (*ent*) Mosquitoes. Insects with mouthparts modified for piercing the skin and sucking blood (females only); wings with scales on the veins and hind margins. They are distributed throughout the world from the tropics to the arctic regions. Eggs are laid in, or near, water (fresh, brackish or stagnant) or in places subject to flooding, according to the species. The larval and pupal stages live in water, both moving actively in the water, coming to the surface frequently to breathe. A blood meal is required by female mosquitoes before they can lay viable eggs.

Mosquitoes cause annoyance by their presence, reducing the amenity value of parks and recreational areas. To some people their bites cause severe irritation and swellings. They transmit diseases of man and domestic animals; malaria is transmitted by species of ANOPHELES (*q.v.*), yellow fever by *Aëdes aegypti* (see AËDES), and species of *Aëdes* and *Culex* are involved in the transmission of some types of encephalitis virus. Species determination of mosquitoes is difficult and the help of a qualified expert is usually necessary.

Cumulative poison. (*tox*) A poison which increases in the body by successive additions, brought about by repeated chronic exposures, the body being unable to detoxify the substance to any degree (e.g. THALLIUM SULPHATE).

Curculionidae. (*ent*) The family of the Coleoptera containing the weevils; insects with an elongated snout (rostrum) and elbowed antennae. See OTIORRHYNCHUS, EUOPHRYUM, SITONA and SITOPHILUS.

Cutaneous. (*tox*) Pertaining to the skin. See DERMAL.

Cuticle. (*ent*) The outer covering of insects composed of numerous small plates (sclerites) joined by intersegmental membranes and secreted by the hypodermis. Composed of three layers the epi-, exo- and endocuticles, often covered on the outer surface by a thin layer of wax protecting the insect from undue loss of water. Modified to form various sensory structures, spines and setae. The cuticle also lines the fore and hind sections of the gut. See INTEGUMENT.

Cyanide. (*chem*) See HYDROGEN CYANIDE and CYMAG.

Cyanogas. (*p. prod*) See CALCIUM CYANIDE.

Cyclodiene compounds. (*chem*) Heavily chlorinated cyclic hydrocarbons with a characteristic molecular structure; all insecticidal, some with high mammalian toxicity, very insoluble in water (< 0.1 ppm) but readily soluble in aromatic solvents (upwards to 50%). Examples are chlordane, dieldrin, endrin and aldrin, but not including DDT and lindane, which although chlorinated hydrocarbons have a different molecular structure.

Cygon. (*p. prod*) See DIMETHOATE.

Cymag. (*p. prod*) A ready-to-use cyanide powder (a mixture of calcium and magnesium cyanide) for gassing rats outdoors; applied to burrows by long-handled spoon or preferably a GAS PUMP (*q.v.*), moisture in the soil releasing hydrogen cyanide gas. Immediately after application the burrows are blocked with soil or turf and 'heeled-in'; treatment is repeated if rats reopen the holes. Special precautionary measures must be followed for the safe use of Cymag (Fig. 40).

D

Damage. Signs of, (*zoo*) Evidence of infestation as shown for example, by woodwork and finished products gnawed by rats (Figs. 23–25); damaged cartons and shredded paper used as nesting material by mice (Fig. 22); excreta of cockroaches and flies on wall surfaces holes in fabrics produced by textile pests; and the blocking of gutters and defacement of buildings with bird droppings (Fig. 21).

A knowledge of the characteristic appearance of pest damage is essential in assessing the location and extent of a pest problem; of diagnosing the problem in the absence of the pest; in formulating the appropriate measures to achieve eradication.

Dark mealworm beetle. (*ent*) See TENEBRIO OBSCURUS.

Dasyphora cyanella. (*ent*) See GREENBOTTLE FLIES.

DDT. (*chem*) One of the earliest synthetic insecticides, discovered in 1939 and introduced by Geigy under the trade names 'Gesarol', 'Guesarol' and 'Neocid'. The first synthetic chlorinated hydrocarbon with a wide spectrum of activity, making a major contribution to insect control during World War II and probably the most widely applied of all insecticides in post-war years. Many insects, e.g. bedbugs, brown dog ticks and fleas may be resistant. Never very effective against cockroaches. Now greatly limited in use, or banned in some countries, because of long residual action, persistence and accumulation of DDT in animal tissues and the environment.

The technical product is a waxy solid, practically insoluble in water, readily soluble in most aromatic and chlorinated solvents. Insecticidal activity combines contact and stomach action.

The acute oral LD_{50} (rat) is 125 mg/kg: stored in body fat, excreted in milk. Many formulations have been used: wettable powders, emulsions, oil sprays and dusts.

DDVP. (*chem*) See DICHLORVOS.

Death watch beetle. (*ent*) See XESTOBIUM RUFOVILLOSUM.

Debris. (*manuf*) Rubbish. The accumulation of waste material in a form conducive to pest infestation; evidence of inattention to good HOUSEKEEPING (*q.v.*) and PREVENTIVE PEST CONTROL (*q.v.*). In food manufacture, small particles of raw materials or manufactured products which accumulate in dead spaces, under machines, in floor crevices, storage areas and transport, providing material for the establishment of insect and rodent infestations.

DEET. (*chem*) See DIETHYL TOLUAMIDE.

Delicia. (*p. prod*) See ALUMINIUM PHOSPHIDE.

Delnav. (*p. prod*) See DIOXATHION.

Density. (*phy*) The weight of a unit volume of a substance, usually expressed as lb/ft^3 or as g/cm^3.

Deratting of ships. (*proc*) A requirement of the World Health Organisation to prevent the re-introduction of Bubonic plague: compulsory regulations to ensure that rat infestations on ships are reduced to a minimum and that ships are treated if the numbers of rats are more than 'negligible'. Port Health Authorities are required to issue 'deratting' or exemption certificates for ships which are treated or found clear on inspection. International Sanitary Regulations permit the use of liquid baits of sodium monofluoroacetate (Fig. 38), as an alternative to hydrogen cyanide fumigation.

Separate from these regulations, many owners prefer to have permanent rodenticidal baits of warfarin established in the holds of their ships to safeguard against reinfestation in ports when cargo is loaded.

Dermacentor variabilis. (*zoo*) Acari: Ixodidae. The American dog tick. Infests woodlands, the larval and nymphal stages feeding on rodents; only the adult is a pest of man. Often brought into homes on dogs, which are particularly susceptible. Most prevalent in the United States during May–August.

This tick transmits spotted fever (a rickettsia) and causes tick paralysis, a non-pathogenic reaction to bites. Regular treatment of dogs and their sleeping quarters is the most effective means of preventing introduction into the home.

Dermal. (*tox*) Appertaining to the skin. A portal of entry of pesticides into the body, their effect usually being much slower than by ingestion or inhalation. The most widely used measure of dermal toxicity is the dermal LD$_{50}$. See LD$_{50}$.

Dermanyssus gallinae. (*zoo*) Acari: Laelaptidae. Poultry red mite. A common parasite of most wild birds and poultry. Heavy infestations reduce the health of chickens and egg-laying. Mites in buildings come from nests of wild birds (sparrows, starlings and pigeons); the mite will attack man.

A medium-sized mite (1 mm long), red after a blood meal, turning black. Eggs are laid in birds' nests; cracks and crevices, under debris; the nymphs and adults feed at night. The life cycle can be completed in 7–8 days; adults survive starvation for many months.

Dermaptera. (*ent*) The Order of insects containing earwigs. Adults, red-brown to black, long and flattened, terminating in a pair of 'forceps' (often curved in the male but straight in the female); the much folded hind wings are covered by short elytra. An unusual feature of the female is that she looks after her eggs *and* the young nymphs until they disperse. Earwigs frequently enter homes from the garden, occasionally in large numbers; new housing estates are often troubled. The adults are nocturnal and seek crevices for harbourage, e.g. under skirtings, beneath stairs and suspended floors, and in cupboards, under sinks and baths. The word earwig derives from the common belief that these insects are liable to make their way into the human ear, but this is rare, the insect then only seeking shelter.

Earwigs are harmless, but the forceps may give a slight nip. Treat where required with dusts and sprays.

Species which may commonly invade buildings are *Forficula auricularia* (Common or European earwig), widely distributed, and *Euborellia annulipes* (Ring-legged earwig) and *Labidura riparia* (Striped earwig) in the U.S.A.

Dermatophagoides pteronyssinus. (*zoo*) Acari: Epidermoptidae. House dust mite. Exceedingly common, the main source of house dust allergen which causes allergic respiratory reactions (e.g. asthma). Feeds on human skin scales; found especially on mattresses where the scales accumulate. In the U.S., *D. farinae* is the major house dust mite.

Not normally infesting stored food, although *D. farinae* may be found on animal feed in Britain.

Dermestes haemorrhoidalis. (*ent*) Coleoptera: Dermestidae. Often found in homes and office buildings, almost certainly originating from birds' nests, the larvae feeding on waste animal material. Both larvae and adults are casual intruders, usually only in small numbers, not causing damage to property. A large beetle (12 mm long) uniformly dark brown. A similar, related species is *D. peruvianus* with which *D. haemorrhoidalis* is often confused.

Dermestes lardarius. (*ent*) Coleoptera: Dermestidae. The Larder, or Bacon beetle. A common pest of factories handling dry animal proteins (bones, pet foods), of larders and shops (where favoured foods include cheese, bacon and ham); usually associated with indifferent standards of hygiene. One of the larger dermestid beetles (6–10 mm long) oval, the front half of the elytra pale, with a transverse band of spots, the rear half darker. Both larvae and adults damage food. Development from egg to adult: 2–3 months (18–25°C), the adult living for 3 months, often hibernating in unheated premises.

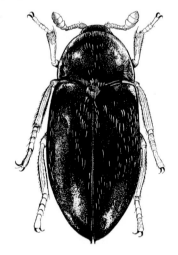

Dermestes maculatus. (*ent*) Coleoptera: Dermestidae. The Leather, or Hide beetle (right). Similar in habits to *D. lardarius*, attacking animal proteins; common in hide and skin warehouses. The adult is uniformly black above, but with white scales on underside of the abdomen; the apex of the elytra produced backwards into a fine point. Larva has a strong tendency to burrow into wood to pupate. Development from egg to adult, 6–7 weeks (23°C).

Dermestes peruvianus. (*ent*) See DERMESTES HAEMORRHOIDALIS.

Dermestidae. (*ent*) The family of the Coleoptera containing many pest beetles of textiles and stored foodstuffs. All with hairy larvae, the majority feeding on materials of high protein content. See ANTHRENUS, ATTAGENUS, DERMESTES and TROGODERMA.

Derris. (*chem*) See ROTENONE.

Desiccant. (*chem*) A dust which abrades or absorbs the outer wax layer of the insect cuticle causing loss of body fluids. The insect then dies by dehydration. See AEROGEL.

Detia GAS EX-B. (*p. prod*) See ALUMINIUM PHOSPHIDE.

Developmental stage. (*ent*) The form of an insect at different periods of its life cycle. See also METAMORPHOSIS.

Diapause. (*ent*) A period of dormancy in insect development which occurs only in certain species and which may take place even when conditions are favourable for continued development. In this latter respect diapause contrasts with HIBERNATION, e.g. in mammals. Diapause may occur in egg, larva, pupa or adult: the best known examples are in larvae about to pupate, lasting for weeks, months or sometimes years.

Some insects enter diapause in every generation—'obligate diapause'; others enter diapause—'facultative diapause'—spontaneously when food deficiency, drought, changes in light intensity, or low temperatures prevail. Not all insects in a population may respond in this way, some completing development in the normal period.

Thus diapause enables some insects to survive adverse conditions which would otherwise be fatal. Examples among pest insects include, *Hofmannophila pseudospretella*, *Ephestia elutella* and *Trogoderma granarium*.

Metabolic activity of insects is often reduced during diapause, with the result that they become difficult to kill with fumigants.

Diatomaceous earth. (*chem*) A powdery silicious rock arising as a deep sea sedimentary deposit of diatoms: minute unicellular plants with flinty shells. Once used as a DESICCANT DUST (*q.v.*). See also AEROGEL. Now commonly used as an absorbent carrier for liquid insecticides in the manufacture of DUSTS and WETTABLE POWDERS.

Diazinon. (*chem*) An organophosphorus compound introduced in 1952 by Geigy, under various trade names, including 'Basudin'. One of the most widely used insecticides in industrial and domestic pest control, especially since the advent of resistance among cockroaches to certain organochlorine compounds. Widely accepted for the control of *Blattella germanica* and *Periplaneta americana*, ants, flies and many household insects. Active also against mites and lice.

The technical product is a pale to dark brown liquid, only slightly soluble in water but miscible with most organic solvents. It decomposes above 120°C, is susceptible to oxidation and slowly hydrolyses in the presence of water to produce a deterioration product highly toxic to man. Commercial formulations are stabilised to prevent this.

Diazinon is highly volatile. Thus when applied to surfaces, the vapour kills flying insects in the air in addition to killing them by contact. It has a relatively slow knockdown and short persistence on absorptive surfaces.

The popularity of diazinon derives from its good insecticidal properties and 'acceptable' mammalian toxicity. It is rapidly degraded in the body and excreted. The acute oral LD_{50} (rat) is 80–150 mg/kg, although recent improvements in the processing procedure have lowered toxicity to an LD_{50} of 300–400 mg/kg; readily absorbed through the skin; inhalation should be prevented in confined areas. Formulations available include emulsion concentrates, oil sprays, wettable powders, dusts and lacquers.

Dibrom. (*p. prod*) See NALED.

Dicapthon. (*chem*) An organophosphorus insecticide with contact action, introduced in 1954 by American Cyanamid Co. A white crystalline powder, insoluble in water, soluble in most organic solvents with useful acaricidal properties and promise for fly control. Used to a limited extent against cockroaches. Of relatively low mammalian toxicity. Acute oral LD_{50} (rat): 330–340 mg/kg.

Dichlorvos. (*chem*) DDVP. An organophosphorus compound with high insecticidal properties first described in 1951 by Ciba; later introduced under the trade names 'Nuvan' (Ciba), and 'Vapona' (Shell). A widely used insecticide for industrial and domestic pest control (cockroaches, ants, bedbugs, fleas, flies and mosquitoes), with contact and stomach action, quick knockdown and useful fumigant properties against flying insects. Often incorporated with other insecticides, instead of pyrethrum, to impart 'flushing action'. A constituent of insecticidal aerosol products and SLOW RELEASE STRIPS (*q.v.*) Non-persistent.

The technical product is a colourless amber liquid with an aromatic odour; slightly soluble in water, miscible with most organic solvents. It is hydrolysed at room temperatures; corrosive to iron and mild steel, but non-corrosive to stainless steel and aluminium.

The acute oral LD_{50} (rat) is 40–60 mg/kg. Formulations available include emulsion concentrates, oil soluble concentrates, aerosols (0·4–1%) and slow release resin strips (20%). Limitations are imposed on the use of the slow release formulations where food is exposed and where people (e.g. in hospitals) may be subject to prolonged inhalation.

Dicofol. (*chem*) A non-systemic acaricide with long residual life introduced in 1955 by Rohm & Haas under the trade name 'Kelthane'. The technical product is a brown viscous oil, insoluble in water, but soluble in most organic solvents. Especially useful for outdoor control of *Bryobia praetiosa* applied as a drenching spray (0·03%) to vegetation; effective against mite eggs and active stages. The acute oral LD_{50} (rat) is 800 mg/kg. Formulations available include wettable powder, emulsions and dust.

Dictyoptera. (*ent*) The Order of insects containing the cockroaches (Suborder Blattaria) and mantids (Suborder Mantodea); insects which produce their eggs in an egg case or oötheca, the ovipositor is much reduced and concealed within the abdomen.

Dieldrin. (*chem*) An organochlorine compound of close chemical affinities with aldrin; the pure compound in Britain is known as HEOD. First introduced as an insecticide in 1948 by Hyman & Co. One of the most widely used insecticides in industrial and domestic pest control with high stomach and contact activity against cockroaches, ants, flies and many other household pests; its value derives from good insecticidal action combined with unusually high stability and long persistence on treated surfaces. Hence its wide use in timber preservatives and soil poisoning for TERMITE CONTROL (*q.v.*).

The technical product is a light tan flaky solid with very low volatility; not soluble in water but soluble in most aromatic solvents.

The acute oral LD_{50} (rat) is 40–50 mg/kg. Readily absorbed through the skin; stored in body fat. Highly toxic to fish. Available in various formulations. Use in many countries is restricted where it is now largely replaced by shorter lived organophosphorus compounds (except for subterranean termite control).

Diethyl toluamide. (*chem*) Deet. A colourless to amber liquid introduced as an insect repellent in 1955 by Hercules Inc. under the trade name 'Metadelphine'; especially effective against mosquitoes and other biting flies on contact.

A viscous liquid, practically insoluble in water but miscible with alcohols, glycols and vegetable oils. Acute oral LD_{50} (rat) 2000 mg/kg: may cause slight skin irritation. Formulated with other repellants (e.g. dimethyl phthalate) in lotions, creams and other bases.

Digestive system. (*zoo*) See ALIMENTARY CANAL.

Diluent. (*chem*) Carrier. Material used to dilute a pesticide, thereby reducing its concentration to the level required for use. Examples are water, oils and SOLVENTS (for sprays), talc and china clay FILLERS (for insecticidal and rodenticidal dusts). The choice of diluent may greatly affect the toxicity of the pesticide through ease of entry into the body, especially the skin. The diluent of many formulations, e.g. oil sprays, may be, orally, more toxic to man than the pesticide, because the diluent is present in vastly greater quantity.

Dimethoate. (*chem*) An organophosphorus compound first introduced as an insecticide in 1956 by the American Cyanamid Co. under the trade name 'Cygon' and by Montecatini by the name 'Rogor'. The technical material forms colourless crystals with a camphor-like smell. Dimethoate is used as a contact and systemic insecticide (in plants and animals) against a range of insects, especially houseflies, cattle grubs and flies of medical importance.

The acute oral LD$_{50}$ (rat) is 320–380 mg/kg. Formulations available include emulsion concentrates and wettable powders.

Dimethyl phthalate. (*chem*) DMP. An insect repellent for personal protection against biting insects, introduced during World War II, after previous use as a plasticiser. A colourless to yellow viscous liquid, with low solubility in water but soluble in most organic liquids. The acute oral LD$_{50}$ (rat) is 8200 mg/kg. Irritating to the eyes and mucous membranes. Applied alone or in combination with other repellents (e.g. DIETHYL TOLUAMIDE) in lotions, creams and other bases.

Dioxathion. (*chem*) An organophosphorus compound introduced in 1954 by Hercules Inc. under the trade name 'Delnav' with non-systemic insecticidal and acaricidal properties, for the treatment of external parasites of livestock. It is used against dog ticks and fleas. The acute oral LD$_{50}$ (rat) is 25–40 mg/kg. Available as an emulsion concentrate.

Dioxacarb. (*chem*) A carbamate insecticide with stomach and contact action, recently introduced by Ciba Limited under the trade name 'Famid', recommended for use against cockroaches and other crawling insects at 1–2% active ingredient.

A white crystalline solid, almost odourless, insoluble in water and odourless kerosene, but soluble in selected organic solvents; unstable in alkaline media. The acute oral LD$_{50}$ (rat) is 100–150 mg/kg; highly toxic to bees. Formulations available include wettable powder (50%) and dust (5%).

Diphacinone. (*chem*) 'Diphacin' (*p. prod*). An anticoagulant of the indanedione type developed as a rodenticide by the Niagra Chem. Div. of Food Machinery & Chem. Corp. A yellow crystalline powder, insoluble in water. A good alternative to warfarin against commensal rats and mice: kills *Rattus norvegicus* in 2–3 days at half the concentration of warfarin, but is slightly less acceptable than warfarin at equal strengths. Somewhat quicker in action than warfarin against *R. rattus*.

Acute oral LD$_{50}$ (rat) 3–17 mg/kg; dog and cat 5–15 mg/kg. Slightly more hazardous than warfarin to domestic animals, although usually used at lower concentrations than warfarin in baits.

Formulations available: an insoluble 0·1% concentrate in cornstarch for making cereal baits (0·005%); a soluble 0·1% concentrate with sugar for making a liquid bait (0·005%).

Diplopoda. (*zoo*) The Class of animals referred to as millipedes from the many legs possessed by some species. In Britain, most have 50–100 pairs of legs, but some have only 17. There are two pairs to each of the majority of segments (cf. CHILOPODA).

Millipedes are narrow cylindrical animals with hard, calcareous skins; they may give off an offensive odour or roll into a spiral when disturbed. Normally live outdoors on damp decaying wood and vegetable debris. Common millipedes fortuitously invading homes in Britain are species of *Cylindroiulus*, brown or blue-grey 15–50 mm long. Eggs are laid in soil; the life cycle may take up to 2 years. Treatment is rarely required.

Diptera. (*ent*) The Order of insects containing the flies; among the most highly specialised of insects. Most with a single pair of membranous wings, and a hind pair of wings modified into small halteres (balancing organs); the mouthparts adapted for sucking, forming a proboscis (as in *Musca*, Fig. 26) or for piercing (as in mosquitoes), the mandibles rudimentary. Larvae are variable in form, all without legs; e.g. the typical 'maggots' of *Musca*, whilst those of *Fannia* resemble plant seeds. The pupa is often enclosed in the last larval skin (puparium).

The Diptera contains many species of importance in pest control; some are just nuisance pests (CLUSTER FLIES), others are of utmost significance to human welfare as vectors of specific pathogens (MOSQUITOES); some are accidental carriers of disease by virtue of their association with excreta, decaying organic matter and food (HOUSE FLIES, BLUE and GREEN BOTTLES). Others are contaminants of food manufacture (FRUIT FLIES).

Dipterex. (*p. prod*) See TRICHLORPHON.

Disinfection. (*proc*) See STERILISATION.

Disinfestation. (*proc*) See ERADICATION.

Dispensing. (*proc*) The act of transferring a pesticide from one container to another for the purpose of dilution, removing a quantity for ease of carrying, or use. It is rarely possible to transfer fluids from containers larger than 1 gall (5 litres) without spillage. This should be recognised in the provision of catch trays, or stillage for very large containers with suitable taps. NEVER dispense pesticides into unlabelled containers or transfer to containers resembling those used for soft drinks or other household items.

Dispersible powder. (*chem*) See WETTABLE POWDER.

Dispersion. (*phy*) A uniform mixture of particles, as in a properly formulated insecticidal dust, or rodenticidal bait. The use of a dye or pigment, first mixed with the active ingredient, helps to give a visual indication of the uniformity of distribution. Also the uniform mixture of two or more liquids.

Disposal, (of empty pesticide containers). (*proc, leg*) The avoidance of environmental contamination and possible risk to children by the safe handling of pesticide containers when emptied. Small amounts of some pesticides are highly toxic to fish. Containers of all sorts have a fascination for children. It is impossible to remove all traces of a pesticide from a container simply by dispensing. The simple rules are: 1) wash out the container after use, 2) dispose of the washings into the sprayer, or nearest drain (not domestic sink), 3) flatten the container (if possible) so that it is unusable, 4) place with refuse for municipal collection or 5)

Fig. 21 (*right*) Bird droppings damaging and defacing the stonework of a window where ledges are used for roosting. These conditions produce unpleasant odour and favourable sites for fly breeding.

Fig. 22 (*below*) Damage by *Mus musculus* in a London warehouse to imported food cartons, resulting in contamination of the contents with urine and faeces.

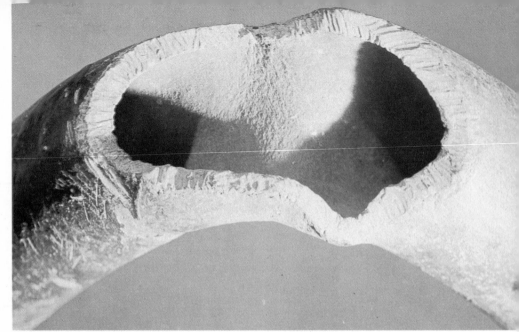

Examples of damage by rats, showing toothmarks:
Fig. 23 (*top*) to lead water pipe;
Fig. 24 (*centre*) to soap;
Fig. 25 (*bottom*) to electric plug.
Gnawing by rats is necessary to keep the incisor teeth short (see Fig. 30). The gnawing of water pipes and electrical installations can result in loss of property and extensive damage.

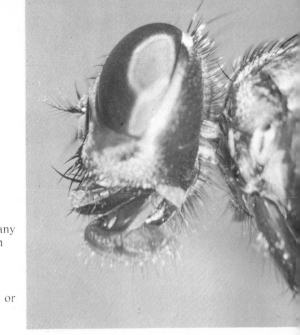

Fig. 26 (*right*) Head of *Musca domestica* showing the proboscis (labellum) for sucking up liquid food. This structure is typical of many Diptera. Bacteria may be carried on its sticky and hairy surface.

Fig. 27 (*below*) *Drosophila*, the fruit or vinegar fly, commonly a pest of canning factories and breweries.

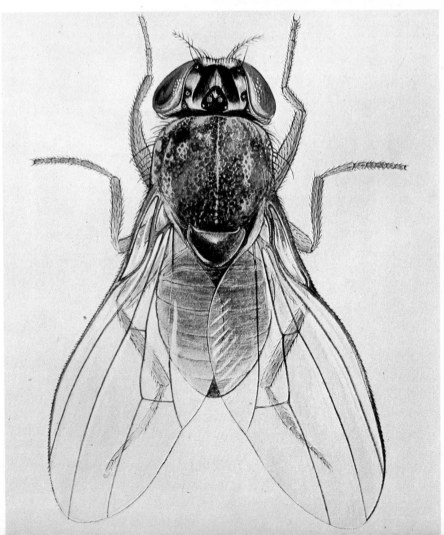

Fig. 28 Use of a dust gun (piston type) to apply insecticide to cockroach harbourages; a common site of infestation where steam pipes pass through a wall.

Fig. 29 Correct carriage of pesticides and equipment in a serviceman's vehicle to prevent spillage and contamination.

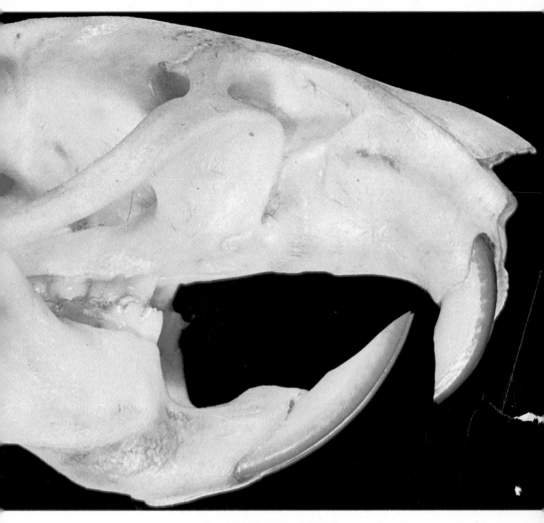

Fig. 30 (*above*) Part of the skull of *Rattus norvegicus* showing the well-developed incisor teeth, kept short by gnawing.

Fig. 31 (*below*) Faeces of *Rattus norvegicus* (left) and *Rattus rattus* (right)

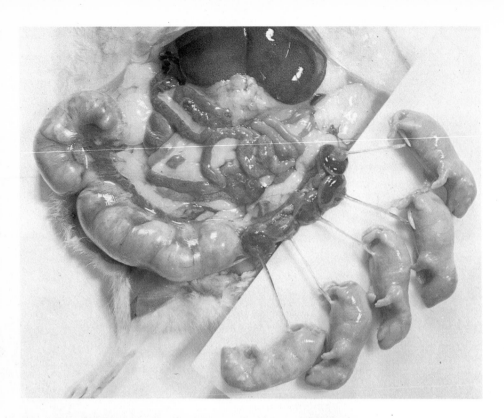

Fig. 32 (*above*) Dissection of a pregnant *Rattus norvegicus*; eac[h] foetus is attached to a placenta b[y] an umbilical cord. More embry[os] are enclosed in the unopened uterus (*left*).

Fig. 33 (*left*) Litter of *Rattus norvegicus* (one week old). Hai[r] just developing but the eyes ar[e] still closed.

Fig. 34 (*above*) Use of a Swingfog machine to control flies on a municipal refuse tip. Note the use of protective clothing.

Fig. 35 (*below*) A hospital kitchen from which food carts (left) are moved to the wards; a means of spreading cockroaches throughout the building.

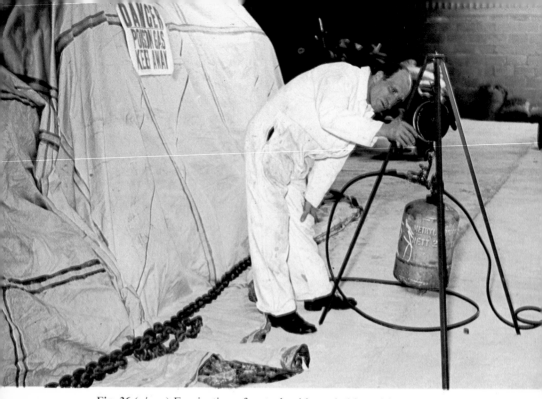

Fig. 36 (*above*) Fumigation of a stack with methyl bromide. Note the use of chains and danger notice.

Fig. 37 (*below*) Fumigation of a building with sulphuryl fluoride; the gas-proof sheets are clipped together where they overlap.

if possible provide for disposal by special arrangement with a local authority. Do not burn aerosol cans.

DMP. (*chem*) See DIMETHYL PHTHALATE.

Dobbin duster. (*equip*) See DUST GUN.

Dog flea. (*ent*) See CTENOCEPHALIDES CANIS.

Domestic. (*name*) A term used by pest control contractors to distinguish a category of clients or source of business, as distinct from commercial/industrial.

Domiciliary. (*zoo*) Pertaining to the home; domiciliary species of cockroach are those closely associated with man, but not only in domestic situations.

Drain. (*bldg*) A pipe or conduit designed to carry to a suitable outfall, waste products that are capable of being removed by the aid of water. The points of difference between a 'drain' and 'SEWER' (*q.v.*) are of administrative importance, in the responsibility for maintenance under the Public Health Act 1936: a drain is a conduit for the removal of waste from one building, or of any structures or yards appurtenant to buildings within the same curtilage (the boundary or ring fence of a single property). Thus, when used solely for private purposes, a drain is private property.

A drainage system should be self-cleansing and free from all liability to cause nuisance, inconvenience, or risk of injury to health; it should be in every sense a sanitary arrangement, that is to say free also, from insect and rodent infestation.

Drat. (*p. prod*) See CHLOROPHACINONE.

Dried currant moth. (*ent*) See EPHESTIA CAUTELLA.

Dried fruit beetles. (*ent*) See CARPOPHILUS.

Dried fruit moth. (*ent*) See EPHESTIA CAUTELLA.

Drie-die. (*p. prod*) See AEROGEL.

Drione. (*p. prod*) See AEROGEL.

Droplet. (*phy*) A small quantity of liquid which remains as an entity in air by virtue of the surface tension at the liquid/air interface. Large droplets settle more quickly than small droplets and coalesce more easily on surfaces. For example, the time required for droplets of water of varying diameter to fall 1·5 m in still air are of the following order:

0·005 mm	30 minutes.
0·01 mm	10 minutes.
0·05 mm	20 seconds.
0·1 mm	5 seconds.
0·2 mm	1 second.

Thus when mists and fogs are used against flying insects it is necessary to have nozzles which provide a small droplet size to give the longest possible time for 'fall out'.

Droppings. (*zoo*) See FAECES.

Drosophila. (*ent*) Diptera: Drosophilidae. Fruit flies, Vinegar flies. Pests of breweries, canning factories and premises where wine, vinegar and pickles are manufactured. Small (3 mm long) yellow-brown, rather bulbous flies (Fig. 27), the abdomen hanging down during flight, which is slow; tendency to hover.

Developmental stages are well adapted for living in fluids and wet fermenting substances, rotting fruit and vegetables. To allow the egg to breathe in liquids it has a filamentous process at one end; the larva has an outgrowth at the end of the abdomen bearing retractile breathing tubes. The pupa has respiratory 'horns'.

Pest species in Britain include: *D. repleta* which breeds in rotting vegetables and is troublesome in kitchens and canteens; *D. funebris* breeds in sour milk.

Fig. 38   Use of sodium monofluoroacetate as a liquid bait for the deratting of ships; treatment is carried out by two men overnight. Note the use of numbered bait positions, non-spill plastic bottle and the use of protective gloves. (Pictorial Press)

Drugstore beetle. (*ent*) See Stegobium paniceum.

Drywood termites. (*ent*) Termites which exist without contact with the soil. Nests constructed in wood, infestations in timber often without visible signs of attack, except expelled frass on surfaces below. Eggs are laid in surface cracks, the young developing within timber; the worker caste is absent.

In contrast to subterranean termites, large chambers are produced within the wood, tunnels are often constructed across the grain. Cellulose is digested with the aid of an intestinal microflora (Protozoa).

The faecal pellets of drywood termites have a characteristic 'poppy seed' appearance. Principal pest species in the U.S. are members of the Kalotermitidae including the widely distributed *Cryptotermes brevis*. Drywood termites are readily distributed in infested furniture. See TERMITE CONTROL.

Duck board. (*manuf*) A slatted structure of wood laid on the floor in wet areas for people to walk on; often found behind bars and in certain areas of food manufacturing premises. Duck boards are undesirable in that they provide numerous crevices for cockroaches; they are rarely cleaned and are unhygienic. They should never be used for storage, in place of pallets, as duck boards give insufficient clearance from the floor for inspection purposes.

Duct. (*bldg*) A channel often forming part of a building structure containing the services, such as plumbing, gas piping and electrical wiring. Ducts may be horizontal or vertical and vary in size considerably: in some hospitals, for example, they form underground passageways, usually containing the central heating pipes, connecting different parts of a building complex. The rough brickwork and warmth provides ideal harbourage for cockroaches (especially *Blatta orientalis* and *Periplaneta americana* and tropical species of ants, especially *Monomorium pharaonis*). Here large populations may develop unnoticed, the interconnecting ducts providing easy access for pests to different parts of a building.

Ducting is frequently created after building construction to hide unsightly pipework. The use of panelling for this purpose is not however recommended, since it provides ideal runs for rodents. If ducting is used, ease of dismantling and inspection should be considered during construction.

Dung beetles. (*ent*) Coleoptera: Scarabaeidae. Insects whose larvae feed on the faeces of various animals; the adults often attracted to light at night, entering premises and causing a nuisance by their presence. A common example in Britain is *Aphodius rufipes*.

Dursban. (*p. prod*) See CHLORPYRIFOS.

Dust. (*chem*) A uniform mixture of a low concentration of an insecticide or rodenticide in a finely divided carrier, usually china clay or talc. Exceptions to this are the use of boric acid as an insecticidal dust (usually 99% active ingredient) or the almost now extinct use of high concentrations of sodium fluoride. In the past, DDT (20–50%) has been used as a rodenticidal contact dust.

The inactive carrier facilitates the uniform application of the active ingredient over the surfaces of the harbourages to be treated. Fine dusts are more effective than coarse dusts (see PARTICLE SIZE); they should only be applied in dry locations since moisture causes the particles to clog (see AGGREGATION). Dusts are more readily picked up by insects than sprays since the particles do not adhere firmly to surfaces. Dusts have the additional advantage of being easily dispersed in cavities not easily reached by sprays, but some have the disadvantage of being repellent to insects. Dusts are the cheapest insecticidal formulation; they are always ready for use and should never be mixed with water. Rodenticidal dusts applied to the harbourages of rats and mice provide a useful control technique in support of baiting. See CONTACT DUST.

Dust gun. (*equip*) Hand-operated equipment for the application of insecticidal dusts and rodenticidal contact dusts, the nozzle often with an extension tube for insertion into cracks and crevices. Principally for use indoors. There are many designs, e.g. piston type ('Dobbin duster') by which dust is expelled by short sharp strokes (Fig. 28): more precise placement of smaller quantities is obtained with bellows and bulb dusters of the squeeze type.

Dust mask. (*equip*) A device covering the mouth and nostrils to protect the wearer against inhalation of particles during the application of insecticidal dusts and rodenticidal contact dusts. The simplest form (Martindale) consists of a light aluminium holder containing a replaceable muslin gauze. During long exposure, wear a light fume mask. See GAS MASK.

Dylox. (*p. prod*) See TRICHLORPHON.

Dynafog. (*equip*) See FOGGING MACHINES.

E

Earwigs. (*ent*) See DERMAPTERA.

Eave. (*bldg*) That part of a sloping roof which at its lowest extends beyond the vertical wall. Gaps left below the eaves, in the facia or soffit boards, provide entry into roof voids of pest birds (notably sparrows), bats (especially in the tropics), and ship rats (*Rattus rattus*). Entry of the latter to domestic properties is facilitated by overhanging trees, electric and telephone cable connections.

Ecdysis. (*ent*) See MOULTING.

Ecology. (*zoo*) The study of the inter-relationship of organisms with their environment. A subject which has come much to the fore in recent years because of possible adverse effects of pesticides in the environment on desirable animals (and plants). A knowledge of the ecology of a pest is necessary to its effective control.

Ectoparasite. (*zoo*) Parasites which feed from the external surface of the host (e.g. bedbugs, fleas and lice).

EDB. (*chem*) See ETHYLENE DIBROMIDE.

EDC. (*chem*) See ETHYLENE DICHLORIDE.

Egg. (*zoo*) The life stage of an organism from which a new individual develops, comprising a zygote (fused nuclei of sperm and ovum, but see also PARTHENOGENESIS); a food reserve (yolk), various membranes (in insects a vitelline membrane and chorion) and often an outer protective calcareous shell (birds).

Egg case. (*ent*) See OÖTHECA.

Electrical hazard. (*equip*) The possibility of being electrocuted when carrying out pest control operations. To reduce this risk, the following should be observed when using powered tools:

1) Use double-insulated tools, or those effectively earthed. A properly grounded tool is the safest way to prevent electrical shock.

2) All electrical equipment should be cleaned and tested at regular intervals by a qualified person. Faults likely to occur are breakdown of the internal insulation and failure of the wiring at its connections. If the tool is not functioning properly, get it repaired immediately.

3) Ensure that tools are fitted with cables appropriate to the rigors of the job, and unaffected by the pest control materials being used. Cables with broken insulation should be replaced immediately. Coil cables neatly when not in use.

4) Wear gloves designed to give protection against commonly occurring voltages.

5) Know the location of live circuits in the area where you are working (especially when drilling), and if possible, turn off the current before starting work.

6) If a fuse 'blows', always replace it by one of the correct value. To reduce hazards when using pesticides in an area containing electrical circuitry (e.g. meter cupboards), use dust formulations in preference to sprays; never use water-based sprays, and avoid the use of metal ladders. See also ELECTROCUTION.

Electrocution. (*phy*) Electrocution occurs when the body becomes part of a 'live' circuit as in one of the following: 1) it comes in contact with both wires (live and neutral) in the circuit, 2) with the live wire and the ground, or 3) with some metal surface (perhaps a faulty electric tool) permitting discharge of electricity through the body to the ground. As a result, the body receives a shock varying in severity with the size of the current, its path through the body, and the length of time that the body is in contact. These influence whether the body feels a mild muscular contraction, possibly leading to paralysis of breathing which can be fatal, or in the extreme case, heart failure. (See ARTIFICIAL RESPIRATION). If the skin is wet, with water or perspiration, more current is allowed to pass and the risk is increased.

The risk of electrocution should be recognised by all those involved in pest control when using powered tools, or when carrying out treatment with pesticides in an area containing electrical equipment, meters, fuse boxes or other wiring. See ELECTRICAL HAZARD.

Elevator. (*manuf, bldg*) Powered equipment for raising commodities usually vertically (cf. CONVEYOR) e.g. of grain in silos, consisting usually of separate 'buckets' attached to a moving belt enclosed in a metal or wooden casing. The commodity is spouted into the elevator at the base and discharged as the buckets are inverted at the top. Insect infestations frequently arise in the elevator boot: the lowest part of the equipment where residues accumulate and high local temperatures are produced by moving bearings.

Also in the U.S., a lift for carrying people; the lift well, usually below ground or basement level is where debris often accumulates to provide fly-breeding sites, rodent harbourage and nesting material. The lift shaft provides ready access to all floors for rodents and insects and the spread of infestation.

Elytron. (*ent*) pl. Elytra. The leathery or sclerotised forewing of an insect serving as a cover for the more membranous hindwings when at rest. In beetles, the elytra are especially horny and meet in a straight line down the back; in cockroaches (where they are called tegmina) they overlap, and in some species are much reduced. In those insects which fly, the elytra are non-functional. Some are modified to provide a stridulatory organ with the metathorax (e.g. *Leucophaea maderae*). In some insects which do not fly, the elytra are fused together (e.g. *Gibbium psylloides*).

Embryo. (*zoo*) The developing organism within an insect egg; in mammals and birds an early stage of development of the foetus (i.e. the period of differentiation before the various parts are visibly recognisable). In man, the foetus before the fourth month of pregnancy.

Emetic. (*tox*) A substance usually given orally to induce vomiting (emptying of the stomach contents by way of the mouth) following accidental swallowing of a pesticide. In cases of emergency, readily available emetics are table salt (one teaspoon) or powdered mustard (quarter teaspoon) to a glass of water and milk (to slow absorption). After each administration, stimulate vomiting by touching the pharynx or back of the tongue with a finger, unless the patient is already vomiting. Apomorphine hydrochloride is the fastest (within 5 minutes) and most effective means of inducing emesis in a conscious patient (0·06 mg/kg body weight). For veterinary use, against suspected poisoning of domestic animals give apomorphine at a dose level of 0·09 mg/kg, preferably subcutaneously. Do not use an emetic if acids, alkalis or kerosene have been swallowed, or if the victim is unconscious.

Tartar emetic has been recommended for inclusion in baits containing acute rodenticides (e.g. Antu) to provide greater protection against poisoning of domestic animals. This is not however recommended as it produces variable reactions and reduces bait acceptance.

Emulsifier. (*chem*) A chemical (e.g. Triton) which allows the admixture of oil and water by stabilising the oil as minute drops within the water. The reverse can also be achieved. Most emulsifiers are synthetic detergents. Soap was once used for this purpose.

Emulsion concentrate. (*chem*) A solution of an insecticide (or other active ingredient usually in the range 20–50%) in solvents, together with emulsifying agents, which allow it to be diluted with water to the concentration required for use. The ready-to-use spray is then milky white in appearance, the opacity being given by the suspension of minute oil droplets (containing the insecticide) throughout the water.

The advantages of an emulsion are: 1) it allows an insecticide insoluble in water, to be applied using water as the carrier, 2) cheaper than ready-to-use oil-based sprays, 3) lower fire risk, 4) less bulk (and weight) to be carried, and 5) applied sprays are less visible on treated surfaces than WETTABLE POWDERS. Disadvantages are 1) ready penetration (in common with OIL SPRAYS) of treated porous surfaces resulting in shorter residual life than wettable powders and 2) possible water staining on susceptible surfaces. Certain emulsion concentrates may be diluted with kerosene or other light mineral oil for use in a fogging machine.

Endemic. (*zoo, dis*) An animal is endemic when habitually present in a particular area. A disease is described as endemic when constantly or generally prevalent in a certain locality due to permanent local causes (e.g. the presence of insect vectors).

Endrin. (*chem*) A highly toxic organochlorine compound little used in industrial pest control and not at all in the U.K. Applications in overseas countries have included outdoor use in mouse control and for the direct killing of birds. Introduced in 1951 by Hyman & Co., an isomer of dieldrin but more toxic than dieldrin and aldrin. Absorbed through the skin; persistent. Acute oral LD_{50} (rat) 10–15 mg/kg.

English Sparrow. (*zoo*) See PASSER DOMESTICUS.

Entanglement. (*phy*) A problem which occasionally arises in the use of a TACTILE REPELLENT of the gel type, when an alighting bird becomes contaminated and unable to fly, possibly causing unfavourable public reaction. Most likely to occur within the first 24 hours of treating a building, particularly with heavy roosting pressure and when the repellent is not adhering sufficiently to the treated ledge. See also STICKY BOARD.

Entex. (*p. prod*) See FENTHION.

Entoleter. (*equip*) A physical method of pest control. A machine, now commonly installed in modern flour mills for the destruction by centrifugal force of insects in milled products (Fig. 79). Effective against *Ephestia kühniella*, *Tribolium confusum*, *Ptinus tectus* and *Acarus siro*, common contaminants of flour and other cereal products.

Environment. (*zoo, chem, phy*) The surroundings in which an organism lives, influencing its growth and survival (see also HABITAT). Of man, factors also influencing his social well-being and aesthetic fulfilment. A word coming much into use in the late 1960's in connection with possible adverse effects of pesticides on wild life. See also PROTECTED SPECIES and TARGET SPECIES.

Enzyme. (*zoo, chem*) A chemical produced by living cells bringing about a specific biochemical change. Highly reactive substances, e.g. amylase in saliva, secreted by some insects, (e.g. flies) over food to pre-digest it; or enzymes of the alimentary tract which reduce food so that it may be absorbed through the gut wall.

Ephestia cautella. (*ent*) Lepidoptera: Pyralidae. Formerly *Cadra*. The Tropical warehouse moth. Also known as the Dried fruit, Dried currant, Almond, or Fig moth. A pest of stored nuts, dried fruit and cereals, commonly contaminating cargoes imported from overseas. It can breed in warehouses in temperate climates during the summer but heavy mortality occurs in winter. The larvae spoil food by their excreta and webbing; in large numbers they can also cause 'souring' of foodstuffs.

The larvae are not unlike those of *Ephestia kühniella*; they feed mainly on

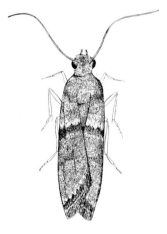

the germ of grain, their webbing often in sheets, covering the exteriors of bagged foodstuffs. The adults fly actively, the wings are grey and banded with lighter colours, not unlike those of *Ephestia elutella*.

Other species of *Ephestia*, notably *E. calidella* are occasionally imported into Britain on shipments of dried fruits.

Ephestia elutella. (*ent*) Lepidoptera: Pyralidae. The Warehouse, Cacao or Cocoa moth, sometimes called Tobacco moth. A general feeder on cereals, pulses, cocoa, dried fruits, nuts and tobacco, occasionally increasing to very large populations. On grain, the larvae tend to concentrate their feeding on the embryos. The adults fly actively. This insect is also a pest of chocolate factories; the possibility of moths from infested beans finding their way to finished products can result in considerable loss of confidence and goodwill of the public. Prosecution may occasionally follow.

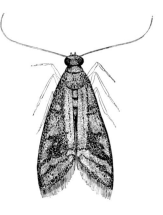

Ephestia kühniella. (*ent*) Lepidoptera: Pyralidae. Previously *Anagasta*. The Mediterranean flour moth or Mill moth, the primary pest of flour mills in temperate and subtropical countries, occurring in the milling machinery, ducts and elevators, particularly abundant in machines processing bran and animal feeds. *E. kühniella* also occurs in warehouses and bakeries to which it is spread by infested sacks. It rarely attacks commodities other than flour.

The larvae are dirty white with a dark brown head and spin a silken tube within which they feed; it is this webbing which blocks the flow of products causing loss

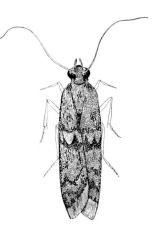

in mill production. The period of larval development (minimum 5 weeks) is much extended by low temperature and foods with poor nutritional content. When mature (12–15 mm long) the larvae wander from the flour to find crevices in which to pupate.

The adult is silver-grey and has a zig-zag pattern of grey and black scales on the wings. It lives for less than two weeks. There are five to eight generations a year in heated mills, but this pest can survive the winter in unheated premises.

Control depends on regular inspection and cleaning, together with the spraying of walls, local treatment of machinery and fumigation (usually once a year).

Epidemic. (*dis*) A disease that affects large numbers of animals or a whole community, in one place in a very short space of time. Spreads rapidly throughout a population resulting in a widespread outbreak (e.g. myxomatosis in rabbits and influenza in humans).

Eradication. (*proc*) The objective of pest control. The removal of the offending organism with whatever safe and economic means available. As a less acceptable alternative, the reduction of the pest to numbers that no longer present a problem. Both objectives may require measures of PREVENTIVE PEST CONTROL (*q.v.*) especially to stop the problem recurring.

Ethohexadiol. (*chem*) Ethyl hexylene glycol, Ethyl hexanediol. An insect repellent for personal protection, effective against most biting insects, first described in 1945. A colourless liquid of faint odour, low solubility in water but miscible in ethanol and related solvents. Not reactive with plastics or clothing. The acute oral LD_{50} (rat) is 2600 mg/kg. Generally formulated with other repellents (e.g. DIMETHYL PHTHALATE).

Ethylene dibromide. (*chem*) EDB. A fumigant with high insecticidal properties first reported in 1925, and introduced under the trade name 'Bromofume' by Dow Chemical Co. in 1946. Its use has been mainly for the control of insects in fruit (developed in Hawaii), as a fumigant for grain and for 'spot' treatment in flour mills.

Ethylene dibromide is a colourless, non-flammable liquid. It is an ingredient of a number of liquid-type grain fumigants but it does not penetrate well into stacks, and needs a longer period of aeration, compared with other fumigants before the vapour is completely dissipated. The gas is six times heavier than air. It has however found extensive use in soil fumigation.

EDB has a strong chloroform-like odour, detectable at about 25 ppm. The acute oral LD_{50} (rat) is 146 mg/kg. EDB causes severe burning of the skin.

Ethylene dichloride. (*chem*) EDC. A colourless liquid with a strong sickly chloroform-like odour, introduced in 1927 as a component of insecticidal fumigants. It is not as toxic to insects as other commonly used fumigants. Because the vapour and liquid are inflammable (flash point 12–15°C), EDC is usually mixed 3:1 by volume with carbon tetrachloride. Such mixtures are commonly used for the fumigation of stored grain, by evaporation of the liquid at 3 gal/1000 bushels (130–225 g/m³). Airing off takes longer than with other fumigants. It is not recommended for use on foods with high oil content.

The acute oral LD_{50} (rat) is 670–890 mg/kg. The odour of EDC is detectable at 50 ppm which gives ample warning of its presence. The toxicity to mammals is relatively low: the danger level to man (30 min. exposure) is 5000 ppm. Damage to the liver and kidneys occurs from excessive doses.

Ethylene oxide. (*chem*) ETO. A fumigant, of intermediate insecticidal activity, introduced in 1928. A mobile, colourless liquid, inflammable in air at concentrations above 3%. Because of the risk of fire and explosion it is normally used admixed 1:9 by weight with carbon dioxide.

The principal application of ethylene oxide has been as a fumigant for bulk grain in recirculating systems and in the vacuum fumigation of packaged foods and tobacco. Fungicidal and bactericidal properties have been exploited in preventing the spoilage of foodstuffs such as spices and the sterilisation of medical supplies and equipment. It also has a lethal action against soil microflora. For insecticidal purposes an application rate to foods of 100 g/m^3 for 3 hours (at 20–25°C) is usually required.

Ethylene oxide produces intense irritation of the eyes and nose which makes it self-warning. The dangerous dose for animals (30 mins.–1 hr. exposure) is 3000 ppm: concentrations above 5000 ppm are lethal after a short time. Contact with the skin may cause burns.

Ethyl formate. (*chem*) A little used insecticidal fumigant; a volatile, colourless inflammable liquid (boiling point 54°C). It is dispensed from automatic machines to individual packages of dried fruits on packaging lines to keep them free from infestation during the short period before they leave the processing plant. In practice ethyl formate diffuses out from the package (depending on the type of pack) in about 48 hours.

The LC (rat) during a 4-hour exposure is 24 mg/litre (8000 ppm).

Ethyl hexanediol. (*chem*) See ETHOHEXADIOL.

Ethyl hexylene glycol. (*chem*) See ETHOHEXADIOL.

Euborellia annulipes. (*ent*) See DERMAPTERA.

Euophryum confine. (*ent*) Coleoptera: Curculionidae. One of a number of wood-boring weevils, including also *E. rufum* and *Pentarthrum huttoni* infesting timbers which are invariably affected by wet rot and therefore usually in damp situations. Sapwood is readily attacked, but heartwood may also be destroyed. Found in 5% of buildings surveyed in Britain between 1960 and 1965, but thought to be increasing rapidly. Usually occur in basements, bathrooms and kitchens but also associated with breweries and wine cellars.

Adults (3–5 mm long) are reddish brown-black with typical weevil appearance. Do not fly. Eggs are laid singly in holes in wood made by the female; hatch in 2–3 weeks, the larvae (3 mm long) are white and legless, maturing in 6–8 months. Tunnelling is similar to *Anobium punctatum*, but most prevalent in the spring wood. Faecal pellets much smaller, darker and more rounded than *A. punctatum*. The pupal chamber is lined with fungal hyphae. Adults emerge during the summer through holes bored at 45° to the surface, about 1·5 mm diameter, irregular in outline and often associated with longitudinal striations on the timber surface. Adults live for about one year, often feeding on the wood (unlike other wood-boring beetles).

European rat flea. (*ent*) See NOSOPSYLLUS FASCIATUS.

European starling. (*zoo*) See STURNUS VULGARIS.

Evidence (of infestation). See DAMAGE, SIGNS OF.

Excreta. (*zoo*) See FAECES.

Excretory system. (*zoo*) The organs of the body providing a regulatory mechanism for the elimination of waste products of metabolism.

Exotic. (*zoo*) Alien. From another country. A pest introduced from overseas, e.g. *Iridomyrmex*, now established in heated buildings in Britain. Cf. INDIGENOUS. Limitation of the import of exotic species is the objective of quarantine.

Exuvium. (*ent*) The cuticle of a larval insect, shed during moulting.

Fabric pests. (*ent*) See TEXTILE PESTS.

Face fly. (*ent*) See MUSCA AUTUMNALIS.

Faeces. (*zoo*) The excreta of animals usually referred to as 'droppings', consisting of indigestible residues of food, bacteria and alimentary secretions, expelled through the anus, their presence in food making it unfit for human consumption. In this respect, the principal offending animals are rodents (*Rattus norvegicus*, *Rattus rattus* and *Mus musculus*), the amount of food fouled by these pests far exceeding the amount they consume. Rats produce an average of 40 droppings per day and mice about 80. Those of the Brown rat are spindle-shaped and usually in groups. Droppings of the Ship rat are sausage-shaped, usually somewhat crescentic, and are more scattered (Fig. 31). Mouse droppings are much smaller in size.

Droppings that are soft and glistening indicate the presence of live rodents; large droppings (from adults) together with smaller droppings (from young) indicate a breeding population. See also BIRD CONTROL.

False ceiling. (*bldg*) A suspended ceiling with a cavity above, constructed for a variety of purposes, often to improve the appearance of rooms with high or apexed ceilings, sometimes solely to allow the use of recessed indirect lighting. The significance in pest control is that false ceilings provide harbourages, especially for mice, and are rarely examined for infestation even if this is practically possible.

False floor. (*bldg*) A floor suspended on joists or battens with a cavity below. Access into such areas provides favourable harbourage for rodent and insect infestation. Inspection covers should be provided whenever possible.

Famid. (*p. prod*) See DIOXACARB.

Fannia canicularis. (*ent*) Diptera: Muscidae. The Lesser house fly (Fig. 43), commonly occurs indoors, especially males, which fly on irregular triangular or square courses in rooms, usually under pendant lamps, cf. MUSCA DOMESTICA. A major pest of poultry houses, causing a nuisance in nearby properties when no effort is made to control them.

The larvae bear irregular protuberances with whip-like hairs enabling them to move through liquid or semi-liquid food (e.g. cow or poultry dung).

Effective control of *Fannia canicularis* depends upon treatment of larvae in breeding sites; poultry house dropping pits; drainage channels outside the houses and piles of poultry manure.

Federal Insecticide, Fungicide and Rodenticide Act (1947). (*leg*) This Act first introduced effective control of pesticides in the United States superceding the Insecticide Act of 1910. The 1947 Act has been amended on a number of occasions, notably in 1959 when it was extended to cover other biocides (nematocides, plant regulators, defoliants and desiccants). Responsibility for enforcement of the Act is now delegated to the Pesticides Regulation Division of the Environmental Protection Agency of the United States. This liaises closely with the Food and Drug Administration in its operation of the Federal Food, Drug and Cosmetic Act (1954) which provides for the control of pesticide residues in raw agricultural commodities.

Feeding behaviour. (*zoo*) The characteristic of animals in the selection and taking of food, and the distance travelled to obtain it. A subject of particular importance in the choice, use and siting of poison baits, especially in rodent control.

Rats and mice eat all types of food (omnivorous) but show a preference for cereals. Ship rats also show a liking for fruits and foods of high moisture content. The amount of food consumed per day varies for the three rodent pests: Brown rat 1 oz (30 g), Ship rat $\frac{1}{2}$ oz (15 g) and House mouse $\frac{1}{10}$ oz (3 g). Rats cut cereals with their teeth giving grain the appearance of being chopped; mice kibble grains, (i.e. remove the outer husk and eat usually only the white endosperm).

Food (together with warmth and harbourage) is one of the major factors which encourage rodents to establish in, or close to, buildings. In this respect, the Ship rat travels more extensively than the Brown rat in foraging for food. The movement of mice between harbourage and food is usually less than 10 metres, often only 3–4 metres, mice sometimes nesting and reproducing within the foodstuff itself. Brown rats require about 2 ozs (60 g) of water per day, Ship rats only $\frac{1}{2}$ oz (15 g) and mice can survive without access to water, providing the foods available have a reasonably high moisture content.

Among most animals there is a hierarchy in the social structure of a population which determines the 'pecking order' in the taking of food. This is pronounced in birds (e.g. the feral pigeon). In addition most animals (rodents and birds) are nervous when feeding and are thus very easily disturbed, birds in particular preferring to feed where their field of view is not restricted. Mice show the strongly developed habit of making quick sorties between harbourage and food, taking small amounts from different sources. This is believed to be an ancestral trait in the avoidance of predators.

Fenchlorphos. (*chem*) An organophosphorus compound known as ronnel in the U.S. and Canada. Introduced in 1954 by Dow Chemical Co. under the trade names 'Nankor', 'Trolene' and 'Korlan'. The technical material is a white crystalline solid which softens at 35°C; insoluble in water but soluble in most organic solvents. Stable at temperatures up to 60°C; incompatible with alkaline materials.

One of the least toxic organophosphorus insecticides to man; used as a residual spray, usually at 1 or 2%, against bedbugs, flies, fleas and mosquitoes. Also against brown dog ticks and cockroaches resistant to organochlorine insecticides, and as a bait for houseflies.

The acute oral LD_{50} (rat) is 1740 mg/kg. This low toxicity also makes Fenchlorphos useful as an animal systemic insecticide. Formulations include an emulsion concentrate for dilution with oil or water.

Fenitrothion. (*chem*) An organophosphorus insecticide, introduced independently in 1959 by Sumitomo Chemical Co. under the trade name 'Sumithion', and by Bayer under the name 'Folithion'; a liquid of low volatility with slight odour, turning brown in contact with alkali. Fenitrothion is a contact insecticide with high activity against cockroaches, and more persistent on treated surfaces than diazinon. In some countries, fenitrothion is permitted for use against insect pests of cereals when applied to the fabric of buildings or boat holds, but not on surfaces against which processed foods may come into contact.

Fenitrothion has an acute oral LD_{50} (rat) of 250 mg/kg; it is not readily absorbed through the skin and does not accumulate in the human body. Dust, emulsion, wettable powder and oil formulations are commercially available.

Fenthion. (*chem*) An organophosphorus compound introduced in 1957 by Bayer under many trade names, including 'Baytex' and 'Entex'. The technical material is a brown oily liquid of low volatility with a weak garlic odour. It has stomach and contact action; used as a residual spray against cockroaches, flies, mosquitoes, and brown dog ticks.

The acute oral LD_{50} (rat) is 215–245 mg/kg but of higher toxicity to dogs and poultry. It readily penetrates the skin. Formulations available include wettable powders, emulsion concentrates, fogging concentrate and granules (for use in mosquito control).

Feral. (*zoo*) A domesticated species which has reverted to the wild state. Particularly of feral pigeons which were originally dove-cote birds.

Feral pigeon. (*zoo*) See COLUMBA LIVIA.

Fertility. (*zoo*) Ability to produce viable offspring.

Ficam 80W. (*p. prod*) See BENDIOCARB.

Field cockroach. (*ent*) See BLATTELLA VAGA.

Field cricket. (*ent*) See CRICKETS.

Field mice. (*zoo*) A collective term for outdoor living mice. The principal species in Britain are the Wood mouse (APODEMUS SYLVATICUS, *q.v.*), and the Harvest mouse (*Micromys minutus*). Neither of these are recognised as pests, although they may occasionally enter buildings for shelter and food (Fig. 7).

Field trial. (*proc*) See TEST METHODS.

Fig moth. (*ent*) See EPHESTIA CAUTELLA.

Filler. (*chem*) A DILUENT (*q.v.*) of dust and other pesticide formulations composed of solids.

Filter flies. (*ent*) See PSYCHODIDAE.

Fire ants. (*ent*) See SOLENOPSIS.

Firebrat. (*ent*) SEE THERMOBIA DOMESTICA.

First aid. (*proc*) The immediate steps to be taken in the event of accidental poisoning with pesticides, whether actual or suspected. They do not replace medical treatment by a qualified practitioner, but should be undertaken while waiting for medical assistance. The seven general principles of first aid are:

1) Send for a doctor and obtain medical aid and advice as soon as possible.
2) Never administer anything orally to an unconscious victim. (See EMETIC.)
3) Loosen all tight clothing, especially round the neck and waist.
4) Keep the victim warm and quiet.
5) Do not leave the victim alone until recovery is complete, or until medical aid has been obtained.
6) Always be prepared to administer ARTIFICIAL RESPIRATION (*q.v.*).
7) In instances of surface contamination, remove contaminated clothing and thoroughly wash affected parts with soap and water.

For certain pesticides there are specific ANTIDOTES (e.g. amyl nitrite for poisoning by HYDROGEN CYANIDE).

Fish moth. (*ent*) See LEPISMA SACCHARINA.

Flat grain beetles. (*ent*) See CRYPTOLESTES.

Fleas. (*ent*) See SIPHONAPTERA.

Flies. (*ent*) See DIPTERA.

Flour mite. (*zoo*) See ACARUS SIRO.

Fluoracetamide. (*chem*) Compound 1081, 'Fluorakil'. An organofluorine compound of high toxicity used as an acute rodenticide in cereal baits. Application in the U.K. is restricted by the Poisons Rules, 1970, to use in sewers, and to drains in restricted areas on port docksides.

The technical material is a crystalline solid, very water soluble, odourless and tasteless. Intensely poisonous to mammals, converted to fluorocitrate in the body, upsetting carbohydrate metabolism. Not so rapid in action as SODIUM MONOFLUOROACETATE (*q.v.*). No known antidotes exist for either of these poisons.

The acute oral LD_{50} (rat) is 15 mg/kg; dogs are much more susceptible. Readily absorbed through cuts and abrasions of the skin. Formulations available include 'Fluorakil 100' (technical material incorporating nigrosine dye) and 'Fluorakil 3' (a 3% cereal bait for admixture, two parts with one of water). Safety is best ensured by storage in a POISONS CUPBOARD (*q.v.*), by restriction of use to experienced personnel, and emphasis on handling with extreme care.

Fluorakil. (*p. prod*) See FLUORACETAMIDE.

Fluorine compounds. (*chem*) Inorganic and organic pesticides containing the element fluorine. See SODIUM FLUORIDE, SODIUM MONOFLUORO-ACETATE and FLUORACETAMIDE.

Flushing agent. (*chem*) A component of cockroach sprays to stimulate the insects to leave their harbourages, thus indicating the extent of the infestation and the areas of a building requiring more intensive treatment. Low concentrations of pyrethrum are commonly used for this purpose; also pyrethroids and dichlorvos. The flushing action of pyrethrum derives from rapid muscular stimulation before paralysis; also possibly its repellent properties to cockroaches.

Fly proofing. (*proc*) See FLY SCREENS, AIR CURTAIN and PLASTIC STRIP CURTAIN.

Fly screens. (*bldg*) A method of fly proofing doors, windows and other ventilation points with mesh: 10 meshes per inch (4 per cm) for flies, 18 meshes per inch (6 per cm) for mosquitoes. Rot-proof nylon is suitable for most purposes, fixed so that doors and windows can be opened; proofing unfortunately reduces ventilation and light. See also AIR CURTAIN and PLASTIC STRIP CURTAIN for entrances involving traffic.

Fly strings. (*chem*) Cords impregnated with insecticide (e.g. diazinon) used for fly control. Vertically hanging objects are favoured by flies as alighting surfaces (hence the success of sticky fly papers), and fly strings provide insecticidal action by contact with alighting flies as well as vapour action in the area treated.

The use of impregnated cords was developed for dairy barns and poultry houses to minimise hazard to animals, before SLOW RELEASE STRIPS (*q.v.*) became available.

Foetus. (*zoo*) The developing young of a mammal in the uterus, attached to the placenta by an umbilicus (Fig. 32), at a stage when its external parts (e.g. limbs) are distinctly formed. Also, an advanced stage of embryological development in the eggs of birds.

Fog. (*chem*) A temporary suspension of droplets, usually of an insecticide, in the air. Fogs are produced by breaking up an insecticide in solution (usually in oil) into minute droplets either mechanically or by heat. See FOGGING MACHINES.

Fogs are particularly effective for the control of flying insects (e.g. *Ephestia elutella*) but have little or no effect on developing larvae. Some insecticidal activity can however be obtained on horizontal surfaces from the 'fall out' of fogs when insecticides with residual action are used (e.g. lindane). There is however little penetration into crevices unless the nozzle of the equipment used is directed into them.

Fogging machines. (*equip*) Equipment for the large scale application of insecticides as a fog (dry) or mist (wet), sometimes also as a spray. Used e.g. for fly control on rubbish tips, mosquito control and for the space treatment of warehouses against moths. Treatment should coincide with the greatest activity of adults and when wind is minimal. Little or no residual action is obtained. In some equipment both water and oil base fluids may be used, but the type of fog obtained varies. Equipment available includes:

Atomisation by hot air blast: air heated by a petrol burner atomises the insecticide in an internal nozzle. E.g. TIFA (Todd Insecticide Fog Applicator): output ranges from 90–180 l/hr; droplet size can be varied to give a dry, medium or wet aerosol. Also suitable for spraying. Usually vehicle-mounted.

Atomisation by exhaust of pulse-jet engine: Portable equipment. E.g. Swingfog (Fig. 34): output ranges from 9–30 l/hr. Dynafog: output 9–270 l/hr. depending on model.

Atomisation by spinning discs: small, portable electrically operated equipment. Small droplets of insecticide are produced by the centrifugal force of spinning discs rotating at high speed. A blower ejects the droplets at high velocity. E.g. Microsol generator: output varies with model.

Atomisation by compressed gas: small portable equipment using gas (nitrogen) at 20–70 kgf/cm² to eject the insecticide through a fine nozzle. E.g. Hi-Fog suitable for fogging, misting and spraying.

Folithion. (*p. prod*) See FENITROTHION.

Food cart. (*equip*) An American term for a mobile insulated metal container (Fig. 35), used in hospitals and other institutions, for carrying hot meals from kitchens to wards; invariably providing harbourage for cockroaches and spreading infestation from one area of a building to another. Insecticidal treatment must ensure that food does not become contaminated.

Forficula auricularia. (*ent*) See DERMAPTERA.

Formaldehyde. (*chem*) A powerful bactericide and fungicide introduced as a disinfectant in 1888, used in soil fumigation and for bacterial sterilisation (e.g. poultry houses) by spray and fog.

A colourless gas with pungent irritating odour and high chemical reactivity. The usual formulation is a 40% aqueous solution (formalin) containing methanol to delay polymerisation.

Formicidae. (*ent*) The family of the Hymenoptera containing the ants; small social insects with the abdomen distinctly constricted behind the thorax to form the pedicel, which may have one or two freely movable knobs or scales. Antennae elbowed. Winged swarming forms are produced seasonally by most species. Often the most persistent of pest insects, the majority living outdoors gaining access to buildings during foraging, a few species living indoors constructing their nests in the fabric of buildings. The majority are nuisance pests, contaminating foods. The number of pest species varies with climate. Those in Britain include, *Lasius niger* (the Black garden ant) and introduced tropical species *Monomorium pharaonis* and *Iridomyrmex humilis*. Additional pest ants of warm climates are the Fire ants (*Solenopsis* spp.), Carpenter ants (*Camponotus* spp.) which can cause serious damage to timber, and various Honey ants. See also under individual species.

Formulation. (*chem*) A convenient form of a pesticide which allows it to be used effectively and safely at a required concentration for a specific purpose. Incorporating a diluent, e.g. oil, water or other solvents (for sprays), fillers and other inert ingredients (for dusts and powders), and surfactants (for emulsions). The ingredients of rodenticidal formulations usually contain a food base, sometimes additives to improve palatability and occasionally binders (e.g. waxes) in block baits.

Fig. 39 Loading crated goods into a fumigation chamber.

Frass. (*ent*) The excreta and food debris of insects, e.g. the faecal pellets of larval textile pests, *Hofmannophila* and *Tineola*, and the damaged fibres which have not been eaten. Equally, the faecal material and particles of uneaten wood produced by the larvae of woodboring beetles (boredust).

Free flowing agent. (*chem*) See ANTICAKING AGENT.

Freon. (*p. prod.*) See PROPELLENT.

Fruit flies. (*ent*) See DROSOPHILA.

Fumarin. (*chem*) Coumafuryl. An anticoagulant rodenticide of the hydroxy-coumarin type, allied to warfarin and equal in efficacy against *Rattus norvegicus* and *R. rattus*, but somewhat less effective, requiring longer feeding, against *Mus musculus*. The technical material is a fine white-creamy powder; chronic oral LD_{50} (rat) 1·4 mg/kg per day for 5 consecutive days; acute oral LD_{50} (rat) 200–400 mg/kg.

Fumigants. (*chem*) Volatile chemicals, vapours entering the body by inhalation; or (in insects) also through the body surface. A miscellaneous group of chemical compounds used as gases to control insects in stored foodstuffs, buildings and ships, and sometimes for rodent control. They may be applied directly as a gas (hydrogen cyanide, sulphuryl fluoride) or as a liquid which vaporises to the gas phase (carbon tetrachloride and ethylene dichloride). Two of the most widely used fumigants for the treatment of foodstuffs are methyl bromide and phosphine (from aluminium phosphide). Certain of the contact insecticides (e.g. lindane and dichlorvos) evaporate readily and are described as having fumigant action. Fumigants have no residual properties following 'airing off'. No fumigant should be used without the appropriate safety equipment and without training or previous experience in fumigation techniques.

Fumigation. (*proc*) The application of a toxic substance in the form of a gas, vapour, volatile liquid or solid in the atmosphere of a closed container (e.g. fumigation chamber, Fig. 39) or structure (e.g. building or stack, under gas-proof sheets) which can be made sufficiently gas tight for the purpose (Figs. 36 and 37). The required concentration of gas must be maintained for a specified period (see CT PRODUCT) and the container or space subsequently ventilated. For some pests (e.g. insects in foodstuffs) fumigation is the only satisfactory method of treatment. The amount of fumigant used (e.g. with methyl bromide) is often dependent on the temperature and the degree of gas-tightness achieved.

Fungal infections. (*dis*) Diseases arising through the activity of fungi. Examples in pest control are the yeast *Cryptococcosis neoformans* and the fungus *Histoplasma capsulatum*. Dust from bird droppings or soil con-taminated with droppings may be infected with these fungi and can be readily inhaled under dry conditions. Cases of Cryptococcosis and Histo-plasmosis (rare) have been recorded among workers in bird infested buildings in the U.S.A. Histoplasmosis can also be contracted from the shovelling or sweeping of bat guano.

Fungus beetles. (*ent*) Beetles of several families including Cryptophagidae, Lathridiidae, and Mycetophagidae, which feed on moulds, sometimes those growing on damp plaster (Fig. 41), hence the alternative name for some species of Plaster beetles. Common in all ill-ventilated buildings with high humidity (cellars, warehouses, flour mills), occasionally contaminating foods. Minor pests in houses shortly after erection, before the plaster has dried out.

Common species are *Anthicus floralis, Cryptophagus acutangulus, Enicmus minutus, Mycetaea hirta* and *Typhaea stercorea.*

Ventilation and drying out of premises eliminates mould growth and the insect infestation. Otherwise control with a mouldicide, with or without insecticide, sprayed onto walls. Destroy infested foodstuffs.

Fungus garden. (*ent*) A globular structure produced within the nest of some termites (the Macrotermitidae) consisting of termite faecal matter (mostly lignin) which supports the growth of certain Basidiomycete fungi. It is believed that these fungi break down the lignin into constituents suitable as food; an alternative to the proctodeal feeding of other termites. See WORKER TERMITE.

Fur beetle. (*ent*) See ATTAGENUS PELLIO.

Furniture mite. (*zoo*) See GLYCYPHAGUS DOMESTICUS.

G

Gallery. (*ent*) An excavation or tunnel produced in timber by a wood-boring insect, such as the larva of a wood-boring beetle, ants (e.g. *Camponotus* species) or termites. Varying in appearance with the feeding behaviour of the insect and often packed with boredust.

Galley. (*ship*) The cooking or food preparation area of a ship, often supporting infestations of *Blattella germanica*, presenting the same problems of control as in commercial kitchens, but with the following additions: 1) Temperatures are often higher because of location in below deck areas; reproduction is increased and insecticides shorter lived. 2) Cooking equipment is confined to a small space, often immovable; difficulties are encountered in treating harbourages with insecticide. 3) Harbourages are often in the cavities of metal bulkheads which cannot be treated without drilling. 4) Inter-connecting ducts and ventilator systems present maximum opportunities for the spread of pests. 5) The opportunity for re-inspection and treatment at port is limited because of normal shipping movements.

Gamma-BHC. (*chem*) Better known by the common name 'lindane', containing not less than 99% of the gamma isomer of benzenehexachloride. Its insecticidal properties were discovered in the early 1940s when the compound was introduced by I.C.I. under the trade name 'Gammexane'.

A colourless crystalline solid only slightly soluble in water, but soluble in most aromatic and chlorinated hydrocarbon oils. Stable to air, light and heat with high insecticidal activity (both stomach and contact) and some fumigant action to a wide range of pests.

The acute oral LD_{50} (rat) is 125 mg/kg; readily detoxified in man. A wide range of formulations are available: wettable powders, emulsion concentrates, dusts, oil-based sprays and smoke generators. The use of lindane, alone or in combination with DDT, in THERMAL VAPORISING UNITS; (*q.v.*) is no longer permitted in domestic properties in certain countries. Use in food storage areas is also restricted.

Lindane has proved valuable for the control of many household and food storage pests, and animal parasites. In recent years lindane has replaced some previous uses of dieldrin especially in timber preservatives. Resistance to lindane exists in bedbugs, brown dog ticks and German cockroaches. A CONTACT DUST of 50% lindane has proved a valuable replacement for DDT for mouse control.

Gamma rays. (*phy*) Electromagnetic (ionising) radiation of penetrating properties which has been considered as a method of sterilising and killing insect pests of foodstuffs, especially grain. The use of gamma rays for this purpose has been proved experimentally but in practice the equipment and safety measures necessary involve high capital expenditure. A regular flow of product for treatment is also necessary to keep the plant fully employed to enable the process to compete economically with fumigation.

Gammexane. (*p. prod*) See GAMMA-BHC.

Garden weevils. (*ent*) See OTIORRHYNCHUS.

Gardona. (*p. prod*) See TETRACHLORVINPHOS.

Gas. (*phy*) A pesticide entirely in vapour form which distributes itself to fill the whole available space, e.g. insecticidal fumigants, some heavier and others lighter than air.

Gas mask. (*equip*) A device covering the mouth, nose and eyes to protect the wearer against inhalation of toxic gases and sometimes sprays and dusts. An essential piece of pest control equipment. Three forms of gas mask are generally recognised:

1) *Light fume mask:* a moulded rubber mouth and nose shield with filters to protect against very low concentrations of vapours from sprays. Also effective against dusts and liquid droplets. This mask does not protect against fogs and fumigants.

2) *Gas respirator:* a full face gas mask to protect against low concentrations of gases (e.g. methyl bromide, hydrogen cyanide, carbon tetrachloride and others) not exceeding 2% by volume in the air. The canister (filter) is interchangeable and usually colour coded to protect the wearer against gases of different chemical composition. Always ensure that masks fit the face correctly before use and that the correct canister is fitted.

3) *Distance Breathing Apparatus:* a full face respirator with fresh air breathing line for protection against high concentrations of gas and solvent vapour. The intake end of the breathing line must be firmly secured in a supply of fresh air to avoid accidental inhalation of toxicant.

Gas-proof sheet. (*equip*) A covering which retains gas for a sufficient time to enable safe and efficient fumigation without moving a commodity from its place of storage, or the vehicle in which it is being transported.

Tarpaulins were once used for this purpose; now largely replaced by lighter weight sheeting of 1) heavy gauge (0·1 mm) polyethylene film, 2) nylon or terylene fabrics coated both sides with neoprene, polyvinyl chloride or butyl rubber, and 3) cotton fabrics coated both sides with neoprene. Usually removed after fumigation to allow aeration; sometimes left in position to protect foodstuffs against reinfestation. To cover large stacks one or more may be joined by rolling one metre of adjacent edges together; clips may be necessary on the junctions of sheets used outdoors (Fig. 37); at least 30 cm of sheeting should be left at the base for sealing to the floor. See SAND SNAKE.

Gas pump. (*equip*) A specially designed pump (Fig. 40), for the application of CYMAG (*q.v.*) to rat burrows outdoors with minimum hazard. The canister and tube must always be emptied carefully after use and before being carried in vehicles. Do not clean by washing.

Gel. (*chem, phy*) A pesticide formulation often used as a bait combining an active ingredient (e.g. chlordecone) in a food base. High viscosity of the gel is produced by the inclusion of a substance which causes a lattice structure of the molecules.

Genetics. (*zoo*) The branch of biology that deals with descent, variation and heredity, and of concern in pest control to those studying resistance to pesticides of insects and rodents.

German cockroach. (*ent*) See BLATTELLA GERMANICA.

Gesarol. (*p. prod*) See DDT.

Gibbium psylloides. (*ent*) Coleoptera: Ptinidae. Hump spider beetle. Not a serious pest of stored products, but sometimes found in stores and warehouses, scavenging on residues. Typical spider beetle appearance, characterised by shining red, strongly convex elytra, which are fused. Slow moving and long-lived.

Gizzard. (*zoo*) The section of the gut of birds, behind the crop, with strongly muscular walls for the physical breakdown of food. Similarly in insects, the proventriculus, which is believed to serve various functions—as a dam, pump or titurating organ. The posterior part of the proventriculus is formed into a valve regulating the flow of crushed food into the mid gut.

Globular spider beetle. (*ent*) See TRIGONOGENIUS GLOBULUS.

Gloves. (*equip*) Essential equipment to prevent the absorption of pesticides through the skin, e.g. when mixing insecticides, during long periods of spraying and when applying certain rodenticides. The conscientious pest control operator will wear them always. PVC gloves are more resistant to certain formulations than rubber and more comfortable to wear; cotton fabric gloves are useless for pest control work.

The exterior of gloves should be washed before they are removed thus preventing contamination of the inside.

Glucochloral. (*chem*) See ALPHACHLORALOSE.

Glue board. (*equip*) See STICKY BOARD.

Glycerine. (*chem*) Glycerol. A viscous liquid which absorbs moisture from the air (a humectant), useful as a binder for baits; also for maintaining baits in a moist condition (e.g. for control of *Rattus rattus*).

Glycyphagus. (*zoo*) Acari: Tyroglyphidae. Mites which occur on many damp foods, and in damp locations, sometimes in newly-built houses before drying out. Common species are *G. domesticus*—the Furniture or House mite—which may cause a skin reaction in sensitive people, and *G. destructor*, frequently in association with the Flour mite (*Acarus siro*). *G. domesticus* may infest upholstered furniture containing vegetable fibres, which readily absorb moisture and support moulds.

Gnathocerus. (*ent*) Coleoptera: Tenebrionidae. Small beetles (4 mm long); pests of minor importance in flour milling machinery, feeding on cereal fractions. The name of this genus is derived from the horny projections on the head of males, and the well-developed mandibles. Two species are common, *G. cornutus*—the Broad-horned flour beetle and *G. maxillosus*—the Slender-horned flour beetle.

Red-brown, slightly larger and more shiny than *Tribolium*. The minimum period of development from egg to adult is about 2 months (27°C). Adults live for about 1 year; unable to survive below 10°C, overwintering occurs in heated premises only.

Golden spider beetle. (*ent*) See NIPTUS HOLOLEUCUS.

Gonads. (*zoo*) The gamete-producing organs: testes of the male (producing sperm), ovaries of the female (producing eggs).

Goodwill. (*manuf*) The established custom or popularity of a business or trade, often appearing as one of its assets with a marketable value. Insects or parts of their bodies, rodent droppings and rodent hairs in food sold to the public, reflect on the image of the company manufacturing and selling the goods and result in loss of custom and goodwill. Adverse publicity resulting from cases of food CONTAMINATION (*q.v.*) can be guarded against by a conscientious approach to pest control.

Gooseberry red spider mite. (*zoo*) See BRYOBIA PRAETIOSA.

Gophacide. (*p. prod*) An organophosphorus compound used as an acute rodenticide to control pocket gophers (species of *Thomomys*, *Geomys* and *Cratogeomys*). Originally introduced by Bayer and developed as Gophacide by Chemagro Corp., undergoing trials against commensal rodents for which it is registered in some countries. Baits contain 0·1–0·5% active ingredient. Pre-baiting is necessary to maximise efficacy.

The technical material is a white crystalline powder, with slight odour, unstable to alkali. The acute oral LD_{50} (rat) is 5 mg/kg. Secondary poisoning is highly unlikely.

Grain heating. (*phy*) An increase in temperature of grain in storage caused by biological organisms. Insects produce heat and water during growth and both are retained by large volumes of grain. These conditions further accelerate growth and reproduction, cause bacteria and moulds to develop, and the grains to germinate. To prevent this, many silos are equipped with temperature measuring equipment to indicate when local heating occurs. The grain may then be turned from one silo bin to another to disperse a local 'hot spot', or preferably treated to kill the infestation.

Grain protectant. (*chem*) An insecticide incorporated into grain to protect it in storage or during shipment. The insecticide most widely accepted for this purpose is malathion (maximum permissible level, 8–10 ppm) applied, 1) as a dust or emulsion to grain on conveyors, before discharge into ships or bins, 2) as a dust in layers about 1 metre apart to grain in bulk, provided that the grain is screened or mixed thoroughly to remove locally high residues before leaving storage. Additionally, 3) to protect against infestation in bag storage, malathion (emulsion or wettable powder) is sprayed onto the surfaces of stacks of raw grains in jute or similar fabric.

Grain or Granary weevil. (*ent*) See SITOPHILUS GRANARIUS.

Granule. (*chem*) A pesticide formulation in which an active ingredient in a solid carrier (often a type of clay) is formed into discrete particles, larger than a dust. Of limited value in pest control; useful outdoors as a means of applying herbicides, soil insecticides and insecticides for mosquito control (See ABATE and FENTHION); providing 'slow release' characteristics, labour saving in application, not wind borne but more expensive than other formulations. Some insecticides are formulated in foodstuffs as granular baits e.g. in cereals for cockroach and ant control and as sugar bait for flies.

Green bottle flies. (*ent*) Diptera. Pest flies of the families Calliphoridae and Muscidae. Most are carrion feeders, attracted to refuse areas and entering buildings to hibernate, (but rarely *Lucilia* spp. (Fig. 42), which are pests of live sheep). Nuisance flies around windows and clustering (with POLLENIA *q.v.*) in roof voids. Pests of slaughter houses.

Adults (9 mm long) are metallic green, varying in hue. Eggs are laid on dead animals and other carrion. Some oviposit in cow dung. Larvae are typical maggots, pupating in dry places. Development time variable. Principal pest species (in U.K.) are *Dasyphora cyanella* and *Lucilia sericata*.

Green lacewing. (*ent*) See CHRYSOPA CARNEA.

Grey-bellied rat. (*zoo*) See RATTUS RATTUS.

Grooming. (*zoo*) The behaviour shown by most animals of cleaning themselves. Rats and mice may live under very dirty conditions, but their habit of self-grooming is very well-developed. It is by this process that rodenticidal contact dusts, adhering to the feet and underparts of the body are transferred to the mouth and ingested. A similar trait is well developed in insects: drawing the antennae and legs through the mandibles and palps by the cockroach, resulting in the ingestion of insecticide; the scissor-like action of the fore and hind legs of the fly, and the drawing of the leg over the membrane of the wing. In birds this behaviour is referred to as preening: a process of rearranging the feathers with the beak.

Ground beetles. (*ent*) See CARABIDAE.

Grub. (*ent*) A non-technical term: strictly the larva of a beetle or hymenopteran, but also the immature stage of other insects in which the larva is short, squat and ponderous in movement. See CATERPILLAR and MAGGOT.

Gryllotalpa. (*ent*) See CRICKETS.

Guesarol. (*p. prod*) See DDT.

Gulls. (*zoo*) See LARUS.

Gully. (*bldg*) An open or covered channel in the floor for the discharge of washing water or processing effluent, often providing harbourage for cockroaches and other pests, especially when ill-maintained.

It is essential that open gulleys be fitted with an effective grating to prevent the entry of solids: an ounce of moist debris around a gulley trap provides sufficient food for many hundreds of insects, including flies. Gulleys should be thoroughly cleaned and disinfected at least weekly; they should be kept in perfect repair and properly trapped to prevent the entry of rats.

Gut. (*zoo*) See ALIMENTARY CANAL.

Habitat. (*zoo*) The place where an organism occurs naturally. Broadly its geographical distribution, but in detail the special locality or environment in which it is found, including the structures and other organisms with which it is associated. As an example, *Anthrenus verbasci* is frequently found in birds' nests. A description of that habitat would include the location (perhaps roof void), the species of bird (perhaps *Passer domesticus*), the structure and contents of the nest (perhaps dried grass and feathers), and the other insects inhabiting the microclimate (perhaps larvae of *Hofmannophila* and *Attagenus*).

Habits. (*zoo*) The characteristic behaviour of animals (see e.g. FEEDING BEHAVIOUR).

Habituation. (*zoo*) The ability of a pest to accustom itself during its life time to a pesticide (e.g. ANTU), demonstrated by the pest being able to survive increasing quantities of the pesticide through the development within the body of increasing amounts of detoxifying substance. Habituation is not transmitted from parent to offspring. Cf. RESISTANCE. See also BAIT SHYNESS (behavioural habituation).

Haemolymph. (*ent*) The blood of insects consisting of plasma, a clear fluid, containing a number of types of cells, the haemocytes. The haemolymph of insects has no oxygen-carrying function. Often the means by which insecticides are transported to the site of action within the insect.

Halide lamp. (*equip*) A leak detector lamp for indicating the presence and concentration of methyl bromide (MeBr) in a treated space. The pale blue flame changes colour to green then to intense blue according to the methyl bromide concentration as follows:

Conc. of MeBr (ppm)	Reaction of flame
0	No reaction.
10	Very faint green tinge at edge of flame.
20	Light green edge to flame.
30	Light green flame.
100	Moderate green.
200	Intense green; blue at edge.
500	Blue green.
1000	Intense blue.

Ventilation should be continued until there is no reaction from the flame.

Hamelin. (*name*) A town (*Hameln* in German) of 50,000 inhabitants, in picturesque hill country, some 40 km south-west of Hanover on the banks of the river Weser (Fig. 44). Now a tourist centre with many old sandstone and half-timbered buildings giving support to the legendary PIED PIPER (*q.v.*).

Harbourage. (*zoo*) The location in which pests spend most of their inactive time, e.g. cockroaches in crevices of furniture and kitchen equipment, wall cavities, electrical switchgear, broken pipe lagging and packing materials (Figs. 46–49); rats in burrows, mice in wall and floor cavities; stored food pests in the crevices of flooring, machinery and packaging materials; and bedbugs in mattress tufts and behind head-boards. Success in insect control depends on locating these harbourages and treating them with effective insecticides. Location is assisted by the use of FLUSHING AGENTS. Harbourages of rodents can be treated with CONTACT DUSTS.

Harvestmen. (*zoo*) Arachnida: Opiliones. Spider-like animals with very long slender legs. Do not spin webs. Occasional intruders into homes where damp conditions exist.

Hemiptera. (*ent*) The Order of insects referred to as bugs, with mouthparts adapted for piercing and sucking. Most are plant feeders. The Order is divided into two suborders: the Homoptera with the forewings uniform in texture, either wholly membranous or opaque, and the Heteroptera with the base of the forewings opaque and the tip membranous. The pest species are represented by three genera of the Heteroptera: viz. *Cimex* containing the bedbug, *Xylocoris* and *Lyctocoris*, insects associated with and predacious on larvae of beetles and moths infesting stored food-stuffs. See CIMEX LECTULARIUS.

HEOD. (*chem*) See DIELDRIN.

Heptachlor. (*chem*) An organochlorine compound, little used in pest control except occasionally against cockroaches and ants, and as an alternative to aldrin and dieldrin in TERMITE CONTROL (*q.v.*). More toxic and volatile than chlordane, with significant vapour action. The technical product is a soft waxy solid. Acute oral LD_{50} (rat) is 90–130 mg/kg. Formulations available include emulsion concentrates, dusts and wettable powders.

Herbicide. (*chem*) A substance with phytotoxic action exploited for its useful properties in weed control. Two types of chemical are used: 'selective herbicides' for their action against perennial, broad-leaved weeds in turf, and 'total herbicides' for the eradication of all vegetation. Herbicides are available in a range of formulations (e.g. sprays and granules), weather conditions at time of treatment often influencing their effectiveness.

Heteroptera. (*ent*) See HEMIPTERA.

HHDN. (*chem*) See ALDRIN.

Hibernation. (*zoo*) Over-wintering in a state of rest. A characteristic of certain mammals; *not* however, commensal rodents which find sufficiently warm conditions in buildings for their activities to continue. Many insects hibernate (usually as larvae, pupae or adults), over-wintering in a crevice, indoors or out, and continuing their development the following year. See DIAPAUSE.

Hide beetle. (*ent*) See DERMESTES MACULATUS.

Hi-Fog. (*equip*) See FOGGING MACHINES.

High frequency sound. (*phy*) Ultrasonics. Air vibrations above the human audible range (greater than 20,000 Hz), claimed to be repellent to commensal rodents, causing them to vacate infested premises. Equipment is commercially available for this purpose, but evidence from tests shows that the effect on rodents is limited and temporary. Additional practical limitations in use arise from the ready absorption and reflection of high frequency sound by thin partition walls and stored commodities. Of little value in practical rodent control and of no value for repelling birds (sparrows in buildings), which do not hear high frequency sound.

Histoplasma capsulatum. (*dis*) See FUNGAL INFECTIONS.

Hofmannophila pseudospretella. (*ent*) Lepi-doptera: Oecophoridae. The Brown house moth, a common pest of fabrics in the home and a wide range of food commodities, rarely developing to large numbers except under favourable conditions of high humidity and lack of disturbance. The larvae are omnivorous and solely responsible for damage; they are regarded as general scavengers commonly found in debris between floor boards, beneath carpets and under-  felts, in cereal spillage of warehouses and food manufacturing premises, and occasionally attacking bulk wheat and bagged flour. They are also common inhabitants of birds' nests and often infest the corks of wine bottles and the cork inlays of wood block flooring.

    The moth runs quickly when disturbed; the wings are bronze-brown, with darker spots on the fore-wings.

Homoptera. (*ent*) See HEMIPTERA.

Honey bee. (*ent*) See APIS MELLIFERA.

Hopper. (*manuf*) A large container narrowing at the base with a valve at the bottom. A hopper acts as a collecting device and temporary store for a commodity before the next stage in processing.

Hornet. (*ent*) See VESPIDAE.

Host. (*zoo*) An organism in which, or upon which, another organism spends part or the whole of its life, and from which it (e.g. a parasite) obtains nourishment or protection. Examples of pests which are host animals are the rat (for the rat flea) and the cockroach (for human hook-worm and dog hookworm).

House borer. (*ent*) See ANOBIUM PUNCTATUM.

House cricket. (*ent*) See CRICKETS.

House dust mite. (*zoo*) See DERMATOPHAGOIDES PTERONYSSINUS.

House fly. (*ent*) See MUSCA DOMESTICA.

Housekeeping. (*manuf*) Maintenance of tidy conditions conducive to effective cleaning and easy inspection for pest infestation. Essential in food manufacturing premises where it is very easy for discarded and redundant materials, left in corners of production and storage areas, to provide rodents with harbourage and nesting materials. Good standards of housekeeping, stacking (Fig. 51), and regular inspection must be practised in all areas of a factory complex: these standards apply as much to laboratories, kitchens, cloakrooms, linen rooms, fitters' shops, boiler houses and offices, as to the production and storage areas.

House longhorn beetle. (*ent*) See HYLOTRUPES BAJULUS.

House mite. (*zoo*) See GLYCYPHAGUS DOMESTICUS.

House mouse. (*zoo*) See MUS MUSCULUS.

House sparrow. (*zoo*) See PASSER DOMESTICUS.

House spider. (*zoo*) See TEGENARIA.

Human flea. (*ent*) See PULEX IRRITANS.

Hump spider beetle. (*ent*) See GIBBIUM PSYLLOIDES.

Hydrocyanic acid. (*chem*) See HYDROGEN CYANIDE.

Hydrogen cyanide. (*chem*) Hydrocyanic acid, HCN. A colourless gas with an odour (to some people) of bitter almonds, used for the fumigation of flour mills, stored grain and ships, its use now largely replaced by METHYL BROMIDE (*q.v.*). HCN is a colourless liquid below 26°C and is supplied compressed in metal cylinders and as absorbed discs. The gas is soluble in water and lighter than air.

Hydrogen cyanide was first used as an insecticidal fumigant in 1886. In the U.K. use became controlled by the Hydrogen Cyanide (Fumigation) Act, 1937, followed by the Hydrogen Cyanide (Fumigation of Buildings) Regulations, 1951, and for marine use, by the Hydrogen Cyanide (Fumigation of Ships) Regulations, 1951. See DERATTING OF SHIPS. It is powerfully toxic to warm-blooded animals: an exposure to 300 ppm for 30 minutes is fatal to man. HCN should not be applied by anyone without specific experience as a fumigator.

In cases of cyanide poisoning, give oxygen and/or ARTIFICIAL RESPIRATION (*q.v.*) and amyl nitrite inhalation. To counteract the rapid poisoning which occurs with HCN, fumigation staff should be equipped with the following freshly made solutions (1 and 2), a glass containing equal quantities, being given to a victim by mouth as first aid:

Solution 1: Ferrous sulphate (158 g) plus citric acid crystals (3 g) dissolved in 1 litre of *cold* distilled water.

Solution 2: Anhydrous sodium carbonate (60 g) dissolved in 1 litre of distilled water.

Seek medical help immediately; usually injection of cobalt edetate.

Hygiene. (*proc*) The science of preserving health by application of the principles of SANITATION (*q.v.*).

Hygiene Officer. (*name*) The person responsible to management, who is actively involved in maintaining the quality and hygiene control of products manufactured in a food industry. His duties include attention to pest control, and pest prevention, either directly, or in close liaison with a pest control contractor.

Hylotrupes bajulus. (*ent*) Coleoptera: Cerambycidae. House longhorn beetle, Italian beetle (in S. Africa), Old house borer (U.S.A.). A widely distributed pest of softwood timbers, but local in distribution in the U.K. Causes serious damage in buildings to structural timbers, which sometimes require extensive replacement.

Adult (7–25 mm long) is dark brown with grey hairs and four grey-white spots (or 2 bands) on the elytra; the pronotum with a pair of shining black eye-like marks and a marked 'waist' between prothorax and elytra (Fig. 53). Adults fly actively in warm weather and live about 3–4 weeks.

Eggs are spindle-shaped, laid in cracks in wood in a number of fan-shaped batches, totalling about 150, and hatch in about 2 weeks. The larvae (30 mm long when fully grown) are white, fleshy with obvious segmentation and small legs (Fig. 52). Larval development varies from 3–10 years. The tunnels are oval in section, often close to the timber surface, occasionally penetrating it to produce longitudinal splits through which the boredust of cylindrical, square-ended pellets is pushed. The surface of the tunnels is transversely ridged (characteristic of many Cerambycids).

Pupation takes 2–3 weeks; adults emerge in July–September (U.K.) making oval exit holes (6–9 mm diameter). Firs, pines and spruce are readily attacked.

Hymenoptera. (*ent*) One of the largest Orders of insects characterised by two pairs of membranous wings, sometimes with reduced venation, coupled together by rows of hooklets. Also by the first segment of the abdomen (the propodeum) being fused and incorporated into the metathorax with (in the majority of families) a constriction in the second abdominal segment forming the 'waist'. The Hymenoptera includes pest species of ants (Formicidae), bees (Apidae) and wasps (Vespidae).

Hypopus. (*zoo*) A resting stage with very short legs occurring under adverse temperatures and humidities between nymphal stages in certain species of mites (e.g. *Acarus siro*).

Hypothermic agent. (*chem*) A substance which interferes with the mechanism of temperature regulation in an animal. A pesticide with this property used for mouse control is ALPHACHLORALOSE (*q.v.*).

Fig. 40 (*right*) Treatment of rat burrows with 'Cymag' applied by a specially designed gas pump. Note the use of full protective clothing.

Fig. 41 (*below*) Mould growth on the damp wall of a cupboard, providing a food source for fungus beetles.

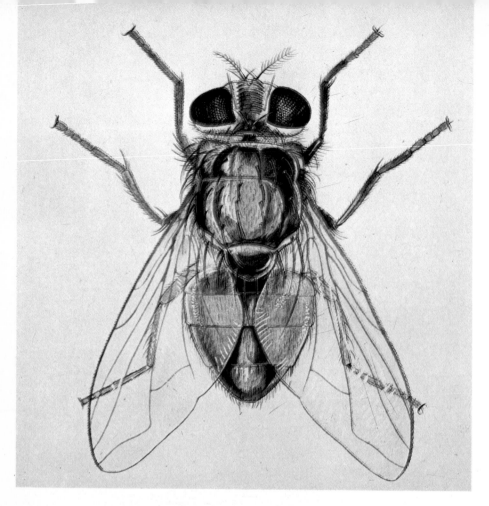

Fig. 42 (*above*) *Lucilia serricata,* primarily a pest of sheep; occasionally entering buildings.

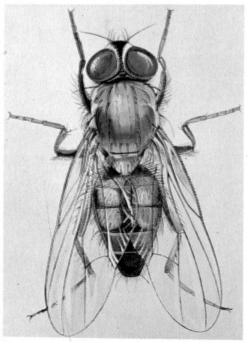

Fig. 43 (*left*) *Fannia canicularis* (male) a major pest of poultry houses causing a nuisance in nearby properties. Males fly on irregular, triangular or square courses below pendant light fittings.

Fig. 44 (*above*) Part of the town of Hamelin with the river Weser: the location of the Pied Piper legend.

Fig. 45 (*below*) Gulls (*Larus* sp.) scavenging on a refuse tip: a hazard to aircraft at international airports. (Nelson)

Cockroach harbourages:
Fig. 46 (*left*) in electric conduit
and wall cavities.

Fig. 47 (*right*) in electrical switch
gear on the wall of a food factory.

Fig. 48 (*left*) in broken insulation
round a steam pipe.

Fig. 49 (*right*) in the corrugations
of packaging material.

Fig. 50 (*above*) An example of a well-kept perimeter area of a food factory, where deliveries are made on a "hard standing" which can be easily washed down. (Marks & Spencer.)

Fig. 51 (*below*) An example of good stacking in a food warehouse, allowing easy access for inspection and rodent baiting. (Marks & Spencer).

Hylotrupes bajulus, the House longhorn beetle:
Fig. 52 (*above*) larva surrounded by bore dust.

Fig. 53 (*below*) adult.

Fig. 54 (*above*) Application of insecticidal lacquer to metal stanchions of kitchen equipment where frequent washing may remove other formulations of insecticide.

Fig. 55 (*right*) Crystals of dieldrin (highly magnified) which are picked up by cockroaches walking over the dried lacquer film. The insecticide is ingested during grooming. Other insecticides which have been used successfully in lacquers include propoxur, diazinon and malathion.

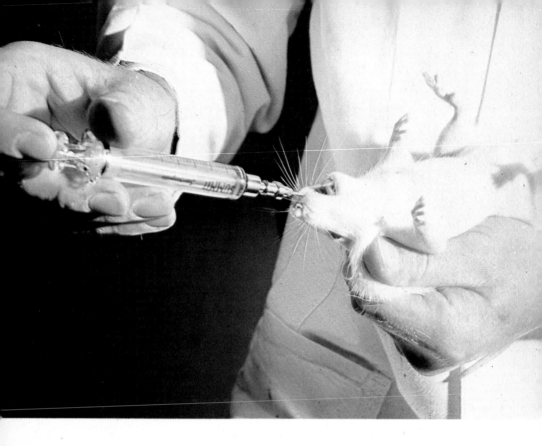

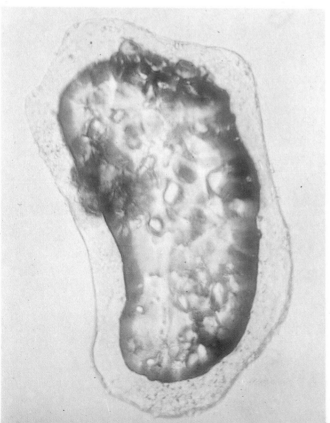

Fig. 56 (*above*) The evaluation of candidate compounds as possible rodenticides by the technique of oral dosing. A measured dose is administered directly to the stomach of a rat by syringe.

Fig. 57 (*left*) Microcapsule of a rodenticidal compound (highly magnified) showing a clear, outer gelatine wall surrounding the active particle.

Imago. (*ent*) See ADULT.

Immunity. (*zoo*) The ability of a pest to withstand the toxic effects of a pesticide: a natural tolerance within a population, or developed by habituation among individuals, or over a series of generations and representing the extreme case of resistance to a pesticide. See also HABITUATION and RESISTANCE.

Impregnated cords. (*chem*) See FLY STRINGS.

Incisors. (*zoo*) The centre pair of teeth in the upper and lower jaw of mammals. Best developed in rodents (Fig. 30) and responsible for the extensive damage to foodstuffs and property; the teeth continue to grow throughout life and must be used continuously to keep their length constant. Hence the propensity of rodents for gnawing.

Incoming materials. (*manuf*) See SUPPLIER.

Incubate. (*proc*) To 'hatch' eggs by means of natural or artificial conditions, often involving the use of controlled humidity and temperature to accelerate the period of embryological development. The hatching time is often referred to as the 'incubation period': also meaning the period between infection with some pathogen and the appearance of symptoms.

Indian meal moth. (*ent*) See PLODIA INTERPUNCTELLA.

Indigenous. (*zoo*) Native to a country, i.e. not introduced (cf. EXOTIC); organisms present in a locality, probably before the advent of man, as distinct from those imported from overseas.

Infection. (*dis*) The entry of a disease organism into an animal or man, by eating infected food, by breathing, or through the skin.

Infestation. (*zoo*) The presence of a pest in a situation where it transmits disease, causes damage, fouling, contamination, fear, offence, or a nuisance, or is potentially capable of having these effects. The term usually implies the existence of a breeding population but may be used to denote the presence of individuals. Defined under the Prevention of Damage by Pests Act, 1949, as 'the presence of rats, mice, insects or mites in numbers or under conditions which involve an immediate or potential risk of substantial loss of or damage to food'.

Inhalation. (*tox*) Intake of a substance by breathing; the route of entry of fumigants and other pesticides with vapour action; a common means of entry into the body of sprays, mists and dusts leading to symptoms of poisoning. See DUST MASK and GAS MASK.

Injector. (*equip*) A device used in the control of subterranean termites to introduce liquid insecticides, under pressure into soil beneath a concrete slab or into the infill of a cavity wall. Often with a mechanism for expanding a gasket at the tip of the injector, producing a seal against the sides of the hole. Used in conjunction with heavy duty hosing to operate safely at the required pressures. See TERMITE CONTROL. Also used in the treatment of large dimension timbers against infestations of wood-boring beetles.

Inorganic insecticide. (*chem*) An insecticide of mineral origin most often used as a dust or bait. The active principle of a number of old formulations, some of which have been brought back into use (e.g. for cockroach control) following the development of resistance to organic insecticides. They include boron (boric acid), sodium fluoride and the more recent silica aerogel. See AEROGEL.

Inorganic rodenticide. (*chem*) A substance of mineral origin, i.e. one not containing carbon compounds (except carbonates), used as a poison for the control of rats and mice. Examples are zinc phosphide, thallium sulphate and yellow phosphorus.

Insecta. (*ent*) The insects, a class of the phylum Arthropoda, whose members are characterised by division of the body into head, thorax and abdomen; with a pair of antennae on the head, a pair of mandibles and two pairs of maxillae, the second pair joined along the centre; also usually with three pairs of legs and in most adults one or two pairs of wings arising from the thorax. Appendages are occasionally borne on the abdomen (e.g. the cerci of cockroaches). METAMORPHOSIS is usual during development.

The Insecta is usually classified into twenty-nine orders, the majority containing pest species.

Insecticidal lacquer. (*chem*) A formulation of a synthetic resin incorporating a high concentration of insecticide, designed for use under conditions where insecticidal dusts and sprays would be quickly removed. Usually applied by brush (Fig. 54). When a lacquer has dried, the insecticide is liberated on the surface, either as fine crystals (e.g. dieldrin, Fig. 55) or as a minute amount of liquid (e.g. malathion). Lacquers are formulated so that the surface deposit of insecticide regenerates whenever it has been removed by abrasion or washing.

Lacquers are effective over longer periods than most other insecticidal formulations, but as only a small amount of insecticide is available on the treated surface at any time, lacquers afford a high measure of safety to man. They are manufactured ready to use, with the exception of those which require an 'accelerator' to be added to speed drying.

Insecticidal smoke. (*chem*) See SMOKE GENERATOR.

Insecticide. (*chem*) A chemical substance which kills insects. The major chemical groups are inorganic, plant extracts, synthetic pyrethroids, organochlorine, organophosphorus, and carbamate compounds. Also a miscellaneous group including methyl bromide, and other fumigants.

In-situ treatment. (*proc*) Post-construction treatment. Remedial treatment. The application of techniques of eradication and preservation; carried out against insect pests of timber at some stage after a building has been constructed. E.g. the surface spraying of roof void timbers with oil-based insecticides against *Anobium punctatum* and the drilling and sub-soil irrigation of buildings against subterranean termites.

Inspection. (*proc*) Survey. The first step to effective pest control; the assessment of the problem, identification of the causative organism(s), leading to the planned action to be taken. Good inspection means good detection work. It involves:

1) Getting the approval and often the assistance of the occupier or property owner.
2) Getting access to all parts of the building and its surrounds.
3) Wearing appropriate clothing to allow you to get dirty.
4) Knowing what to look for and where.
5) Using a good torch to find the evidence.
6) Getting on hands and knees to examine difficult locations.
7) Locating the infestation to its full extent.
8) Asking yourself how the problem arose.
9) Recording what is found.
10) Assessing what needs to be done by way of treatment, and possibly co-operation by the client, in improved sanitation and proofing.
11) Preparing a detailed report incorporating recommendations for action.

Instar. (*ent*) Stage. Stadium (archaic). Widely held to be the form assumed by an insect between successive moults. Recently redefined, more precisely, as the form of an insect between successive apolyses; (the term apolysis referring to the detachment of the epidermal cells from the old cuticle, usually followed by ecdysis, the act of shedding the old cuticle.)

Instructions for use. (*tox*) See LABELLING.

Integument. (*ent*) The body covering of insects consisting of three layers: 1) the outer CUTICLE, largely composed of chitin, 2) the hypodermis, a continuous layer of cells which secretes the cuticle, and 3) the basement membrane, a continuous structure-less layer attached to the inner surface of the hypodermis.

Iodofenphos. (*chem*) An organophosphorus compound, closely related to BROMOPHOS, with insecticidal and acaricidal properties, recently introduced by Ciba Geigy, under the trade name 'Nuvanol N'. Recommended for use against mosquitoes and flies at 1–2% (in dairy barns and slaughterhouses), poultry red mite, flour mite and stored product pests. Especially effective against cockroaches.

The technical product is a colourless crystalline solid with a mild smell, insoluble in water, soluble in many organic solvents, unstable on alkaline surfaces. Of low toxicity: acute oral LD_{50} (rat), 2500 mg/kg, toxic to bees. Formulations available include wettable powder (50%), emulsion concentrate (20%) and dust (5%).

Iridomyrmex humilis. (*ent*) Hymenoptera: Formicidae. Argentine ant. A tropical species which inhabits heated premises, particularly those with high humidity, e.g. certain areas of hospitals, restaurants, laundries. Nests are made in brickwork, hollow spaces in walls, behind stoves, often in inaccessible locations leading to the establishment of large colonies. In warm climates, they may nest outdoors (under buildings, pavements, where the soil is damp). There may be several queens to a nest. Because of extensive wandering by worker ants, they may be fortuitous carriers of pathogenic bacteria in hospitals. The ants are attracted to sweet foods, proteins and fats and exhibit the characteristic of many ants of moving along trails.

The workers are brown (2 mm long), the pedicel with one knob. Development from egg to adult takes about 3 months. In centrally heated buildings, winged sexual forms may appear throughout the year and new nests may be started at any time.

Isecon. (*p. prod*) See PROPELLENT.

Isobornyl thiocyanoacetate. (*chem*) Isobornyl thiocyanatoacetate (in U.S.A.). A contact insecticide causing rapid knockdown of flies, used mainly in domestic fly sprays, the compound introduced by Hercules Inc. under the trade name 'Thanite'. The technical product is an oily yellow liquid, insoluble in water, but miscible with oils and organic solvents. It has an odour similar to turpentine and contains about 20% of related terpene esters.

It is now little used, but has served as a contact insecticide alone, or more usually in combination with other insecticides to replace and synergise pyrethrins. Low mammalian toxicity; the acute oral LD_{50} (rat) is 1600 mg/kg: slightly irritating to the skin, must be kept away from the eyes and mucous membranes. Available in kerosene and other oil solutions and emulsions. Thanite in space sprays is not permitted for use in dairy barns, milk rooms and poultry houses in the U.S.A.

Isomer. (*chem*) A compound composed of the same number and type of atoms as another, but differing in the spatial distribution of the atoms. Some isomers in their atomic arrangement are mirror images of others. The biological activity of different isomers may vary considerably (e.g. *gamma*- cf. *beta*-BHC).

Isopoda. (*zoo*) An Order of the Class Crustacea which includes the woodlice, or sowbugs, the only truly terrestrial crustaceans. Three common species occur in England: *Oniscus asellus* (15 mm long), *Porcellio scaber* (17 mm) and *Armadillidium vulgare*, 18 mm—a species which rolls up into a tight ball when disturbed, hence the name Pillbug.

The thorax has seven segments each with a pair of walking legs. The arched bodies are oval when seen from above, generally grey-black. Woodlice are always found in cool, damp, dark situations. They feed on decayed wood and other vegetation and may be attracted to damp locations in kitchens and bathrooms (around sinks), but are harmless indoors.

Isoptera. (*ent*) The Order of insects containing the termites (white ants). Social insects with biting mouthparts, living in communities of a few dozen to several million individuals, composed of a number of CASTES (*q.v.*). Primarily feeders on cellulose and major pests of timber in the tropics and sub-tropics. When present, the fore and hind wings are similar, elongate and membranous, superimposed when at rest and capable of being shed by fracture at the wing base.

Colonies are formed following dispersal of the alate reproductives (Fig. 15); pairing usually takes place on the ground, but actual copulation may be delayed for several days. Soil-nesting species produce a small cavity in the soil in which the first eggs are laid. The fertilised female, the 'queen' of the new colony, may continue to produce progeny for many years. Workers (Fig. 16), the first members of the colony and ultimately the most numerous individuals are responsible for food gathering, the construction of FUNGUS GARDENS (*q.v.*), the tending of young and building the termitarium. Soldiers (Fig. 14) defend the nest, and fertile male and female alates (primary reproductives) ultimately establish new colonies. These castes are fed by the workers with predigested wood.

In nature, termites convert dead wood to humus but structural timbers in buildings are equally prone to attack. Control of termites is therefore a major pest-control activity in most warm climates. Two types of termite are distinguished 1) SUBTERRANEAN TERMITES and 2) DRYWOOD TERMITES, requiring different techniques for effective control.

Italian beetle. (*ent*) See HYLOTRUPES BAJULUS.

Ixodes ricinus. (*zoo*) Acari: Ixodidae. Common sheep tick, Castor bean tick. So called because of its resemblance to a castor bean seed; occasionally attaches itself to dogs and humans, often brought indoors in rural areas.

Adult (4–5 mm long), grey oval with a hard plate on the anterior region of the back. Eggs are laid among grass roots and newly hatched larvae climb vegetation to attach themselves to a suitable host to obtain a blood meal. This is repeated by the nymphs and adults, all stages being highly resistant to starvation. Development from egg to adult varies widely with the availability of hosts. The bite is not serious to man provided the tick is removed complete with mouthparts.

J

Japanese cockroach. (*ent*) See PERIPLANETA JAPONICA.
Jaundice. (*dis*) See LEPTOSPIROSIS.

K

Kelthane. (*p. prod*) See DICOFOL.

Kepone. (*p. prod*) See CHLORDECONE.

Khapra beetle. (*ent*) See TROGODERMA GRANARIUM.

Knapsack sprayer. (*equip*) A portable sprayer carried on the back of the operator, powered by a petrol engine (or fuelled and lubricated by a petrol/oil mixture) for application of insecticidal sprays over large areas (e.g. fly control on refuse tips). Output varies with model but 2·5 litres/minute is typical. An additional attachment, makes such equipment suitable for the application of insecticidal dusts (Fig. 59).

Knockdown. (*chem*) KD. Incapacitation of an insect by a quick-acting insecticide such as pyrethrum, allethrin or tetramethrin, known as knockdown agents, often incorporated into insecticidal mixtures (e.g. for fly control), with the express purpose of producing rapid paralysis of the insect.

Korlan. (*p. prod*) See FENCHLORPHOS.

Label. (*leg*) Written, printed or graphic matter on, or attached to, a poison (or device) or its immediate container, and the outside container or wrapper of the retail package, should there be one. (As defined under the FEDERAL INSECTICIDE, FUNGICIDE AND RODENTICIDE ACT (1947), *q.v.*, of the U.S.A.)

Labelling. (*leg*) Provision of adequate information on containers used for pesticides: a necessary precaution in the attempt to prevent incorrect use and consequent accidental poisoning. Provisions governing the content of labels are therefore generally included in pesticide control legislation. If read, the label provides the most important information to guarantee the safety of the user of the pesticide.

The requirements vary widely from country to country, but the following are generally included: name and address of manufacturer; statement of active ingredient(s) by correct chemical name; concentration(s) by weight; instructions for use and the precautionary measures to prevent injury to man. In some instances 'standard phrases' have been promoted to describe these measures. Some countries divide pesticides into classes according to toxicity, with a class coding symbol, and different labelling requirements for each. With certain products it is required that the label states early symptoms of poisoning, first aid measures and antidote, if any, and flammability.

In the U.S.A., labelling is defined to include not only labels, as such, but all other written, printed or graphic matter accompanying the poison (or device) or to which reference is made on the label or in accompanying literature. See also DISPENSING.

Labidura riparia. (*ent*) See DERMAPTERA.

Lacewing fly. (*ent*) See CHRYSOPA CARNEA.

Lacquer. (*chem*) See INSECTICIDAL LACQUER.

Ladders. Use of, (*equip*) Accidents arise from the use of ladders because:
1) the ladder is inadequate for the job (too short), or in disrepair and therefore unsafe;
2) the ladder is not supported correctly or is used in an unsafe manner;
3) the ladder leads the man onto an unsafe area.

For heights up to 10 metres the minimum length of ladder should be the height plus 1 metre. Heights from 10 to 13 metres, require an additional 1·5 m of ladder. Extension ladders require a minimum overlap of 1 and 1·5 m, for heights up to 10 m and 13 m, respectively. For maximum safety, the base of a ladder should be one metre away from the wall for every four metres of the height being reached.

Do not stand higher than the third rung from the top and never leave a ladder unattended. Do not take risks by stepping off the top of a ladder onto an 'unknown area'. Talk with the maintenance engineer of the building first. He will know whether it is safe to do what you propose.

Ladybird beetles. (*ent*) See COCCINELLIDAE.

Lance. (*equip*) Part of a piece of spraying equipment; the metal tubing beyond the trigger and carrying the spray nozzle, the length determined by the area to be sprayed and accessibility.

Larder beetle. (*ent*) See DERMESTES LARDARIUS.

Larus. (*zoo*) Gulls. Occasionally pests, notably at airports where birds scavenge off refuse tips nearby (Fig. 45), and use the grassed areas as resting sites. Here they endanger air traffic by the possibility of damage to the structure of planes, the cockpit windows and jet air intakes. See AVITROL and BIRD SCARING DEVICES.

Larva. (*ent*) The immature stage of an insect with a complete metamorphosis. A larva hatches from an egg, and when mature changes into a PUPA (*q.v.*). Characterised by having a form (and often a habitat) totally different from that of the adult (Fig. 4). The wings are never visible, cf. NYMPH.

Larvae are the most damaging stage of many insects of economic importance (e.g. those attacking stored foodstuffs and textiles); most have well developed mandibles. Larvae often live entirely within the foodstuff or within individual grains (e.g. grain weevil). They usually have three pairs of thoracic legs, but some have extra 'false feet' on the abdomen (moths), and some are legless (especially flies). The distribution of thoracic and abdominal setae is often an important key to species identification. In the Acari (mites), a *larva* with three pairs of legs hatches from the egg and quickly moults to the first *nymphal* stage having four pairs of legs.

Larvicide. (*chem*) A chemical with particular activity against larval stages of insects, or used in such a way as to bring about the control of insect larvae (e.g. mosquitoes).

Lasioderma serricorne. (*ent*) Coleoptera: Anobiidae. The Cigarette beetle. The adult (2–3 mm long) is red-brown, densely covered with short yellow hairs; there are no marks on the elytra (cf. STEGOBIUM PANICEUM) and the top of the prothorax is more evenly rounded (cf. the humped thorax of ANOBIUM PUNCTATUM). See illustration on page 20.

The larva of *Lasioderma* is a primary pest of leaf tobacco, cigars and cigarettes and is common in bonded warehouses where fumigation is the principal method of control. The adults do not feed. Infestations in processing machinery may give rise to insect contamination of packaged cigarettes, eggs developing to adults after despatch. Refrigeration, although costly, has been used to check this problem.

Lasius. (*ent*) Hymenoptera: Formicidae. *Lasius* (*Acanthomyops*) *niger*, the

Black garden ant, is indigenous to Britain and common around homes, the workers entering through cracks in brickwork and around windows in search of food. It causes annoyance when found feeding on sweet substances, jams and other preserves in larders and pantries, causing the housewife to reject foods she finds infested. It is not known to carry disease.

Around buildings, the nest openings may be easily located by the presence of small piles of fine earth; simultaneous swarming flights take place over a wide area usually in July–August when mating takes place in the air.

Lasius brunneus may also enter houses, causing annoyance in the same way.

Latheticus oryzae. (*ent*) Coleoptera: Tenebrionidae. The Long-headed flour beetle. A small beetle, about 3 mm long, with an elongate head and clubbed antennae, which infests rice flour and other cereals. Its presence in the U.K. arises from successive importations of infested cargoes from the tropics.

Latrodectus mactans. (*zoo*) Arachnida: Araneae. The Black widow spider. The most common and most dangerous of widow spiders in the United States, having a very potent venom. Usually found in outdoor toilets, garages, outbuildings and drain culverts, hanging upside down in its loosely woven, irregular web. The female's body (15 mm long) is jet black with a red mark, shaped like an hour glass, on the underside. The male is much smaller (4 mm), but both sexes bite.

LD_{50}. (*tox*) LC_{50}. The lethal dose (LD) or concentration (LC) of a chemical that kills 50% of a population (minimum 10) of test animals. The oral LD_{50} is obtained by *feeding* carefully measured doses. The dermal LD_{50} is obtained by applying carefully measured doses to the *skin*. The terms 'acute' or 'chronic' LD_{50} denote whether the treatment was given as one dose, or spread over a period of time. LD_{50} values are expressed as the weight of the chemical in its technical form (usually in mg) per kilogram of body weight of the test animals to bring about 50% kill. To assess the toxicity of a pesticide to man, the laboratory rat is usually used as the test animal. LD_{50} values are often used to compare the toxicities of different compounds, although the LD_{90} or LD_{99} (dose for 90% or 99% kill) may be a more useful indication of the practical effectiveness of a chemical for pesticidal use.

LD_{90}. (*tox*) See LD_{50}.

Leather beetle. (*ent*) See DERMESTES MACULATUS.

Legislation. (*leg*) The governmental process of safeguarding the public, users of pesticides, wildlife and the environment against adverse effects of pest control chemicals. The need to control pesticides is a consequence of the fact that most are poisonous to a varying extent to man and domestic animals, as well as to the pests they are intended to destroy.

Legislation on the control of pesticides is inevitably complex because of the multiplicity of ways in which pesticides can directly or indirectly affect human health. The procedure varies in different countries; it usually involves the authorities responsible for public health, labour and agriculture, and sometimes transport, with the support of various boards or committees in an expert advisory function.

Legislation on pesticides is invariably concerned with the registration of products, licence to manufacture, supply or use, and residues in food-stuffs. There may also be legislation on specialised subjects such as fumigation. The U.K. occupies a special position at the present time; pesticide control is largely a voluntary matter under the PESTICIDES SAFETY PRECAUTIONS SCHEME (*q.v.*). See APPROVAL OF PESTICIDES.

Lepidoptera. (*ent*) An Order containing about 200,000 species, insects with two pairs of membranous wings covered with scales—the butter-flies and moths. The larvae of some moths are serious economic pests damaging cereal products, dried fruits, nuts and animal fibres, many cover the surfaces over which they walk with fine threads of silk (webbing) and spin silken cocoons in which to pupate.

The presence of pro-legs, usually five pairs, on the abdominal seg-ments of the larva is characteristic. The mouthparts of adults are modi-fied into a proboscis (in butterflies for the sucking of nectar) or reduced and non-functional as in moths associated with stored foods. See e.g. EPHESTIA, PLODIA and HOFMANNOPHILA.

Lepisma saccharina. (*ent*) Thysanura: Lepismatidae. Silverfish or Fish moth. A primitive wingless insect, active at night, darting to cover when disturbed; readily distinguished by the slender grey body, long antennae and long bristles on the end of the abdomen. An un-pleasant inhabitant of homes where it prefers warm damp situations in kitchens and pantries, feeding mainly on carbo-hydrates (see right).

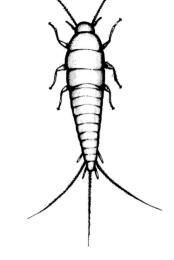

Leptocoris trivittatus. (*ent*) Heteroptera: Leptocoridae. Boxelder bug. A plant-feeding bug often entering buildings as a casual intruder in search of shelter in spring and autumn in many parts of the U.S.

The adults are black (12 mm long) with red markings on the thorax and wings. Eggs are laid in crevices of the bark of female Boxelder trees (*Acer negundo*); the young nymphs are wholly red, the older nymphs, red with black markings. Young and adults feed on the seeds, 'keys' of the Boxelder tree during summer.

The adult hibernates in protected situations in winter, entering buildings when the weather becomes cold. The bugs may stain fabrics indoors and emit a foul odour when crushed; do not feed on stored foods.

Control measures consist of removing female Boxelder trees, treating trees using powered spraying equipment (carbaryl and dipterex are recommended), and treating with insecticidal dusts the surfaces indoors over which the insects crawl.

Leptospira icterohaemorrhagiae. (*dis*) See LEPTOSPIROSIS.

Leptospirosis. (*dis*) Leptospiral jaundice, Weil's disease. A disease spread by rats to humans through contact with rat urine or mud and water infected with the causative bacterium, a spirochaete, *Leptospira ictero-haemorrhagiae.* Up to 40% of rats taken in surveys in the United Kingdom have been shown to be infected. Leptospirosis also occurs in dogs, pigs and calves.

Lesser grain borer. (*ent*) See RHIZOPERTHA DOMINICA.

Lesser house fly. (*ent*) See FANNIA CANICULARIS.

Lesser mealworm beetle. (*ent*) See ALPHITOBIUS DIAPERINUS.

Lethal dose. (*tox*) LD. Lethal concentration (LC). The amount of a substance which kills at least one animal in a test group and thus an imprecise measurement of the toxicity of a chemical. See LD_{50}.

Lethal time. (*tox*) The time required for a toxic substance to kill a test subject. Often expressed as LT_{50}: the time necessary for a given dose of a toxicant to kill 50% of a statistically significant group of test organisms.

Lethane 384. (*p. prod*) The trade mark of Rohm & Haas Co. for an organic thiocyanate insecticide introduced in 1932. A brownish oil, compatible with pyrethrins, piperonyl butoxide, chlorinated hydrocarbon and organophosphorus insecticides, used to improve knockdown. For this purpose it is incorporated into oil-based household sprays and aerosols for use against houseflies, mosquitoes and domestic insect pests. Acute oral LD_{50} (rat) 90 mg/kg. Formulations of Lethane 384 include 50% v/v in kerosene.

Leucophaea maderae. (*ent*) Dictyoptera: Blattidae. The Madeira cockroach. A pest insect in the West Indies, South America, occasionally in the U.S.A., part of Southern Europe, Africa and the East Indies. Favours tropical fruits especially bananas, grapes; occurs in warehouses and fruit stores, occasionally in homes.

A large cockroach (40–50 mm long), the tegmina fully developed, mottled, grey with two curved brown lines in the basal area: slow moving but flies actively. The oötheca is incubated internally by the female, from which the young nymphs emerge. Incubation period 2 months; nymphal development period 4–5 months (30–36°C).

Life cycle. (*zoo*) The period between fertilisation of an egg and death of the adult which develops from that egg. Also, more loosely, the time between egg-laying of an adult and the egg-laying of its progeny.

Life history. (*zoo*) A description, in chronological order of the activities of all stages of development in the life cycle of an organism, during one generation, or series of generations.

Lift. (*bldg*) See ELEVATOR.

Lighter. (*ship*) See BARGE.

Limothrips cerealium. (*ent*) Corn thrips, 'thunder bug'. One of the smallest insects (1·5 mm long), black with very narrow hairy wings. Feed on the sap of grasses, especially wheat. On sultry summer days the insects fly and drift with the wind, often entering buildings in large numbers. Cause nuisance by irritating human skin as they crawl over hairs; also by collecting behind the glass of pictures.

Lindane. (*chem*) See GAMMA-BHC.

Lipase-free. (*chem*) Lacking the enzyme lipase which breaks up fats. Most ready-to-use rodenticides are formulated in a cereal base (e.g. oatmeal) in which the lipase is rendered inactive by heat treatment so as to prevent the bait going rancid in storage and use.

Liphadione. (*p. prod*) See CHLOROPHACINONE.

Litter. (*manuf, zoo*) Miscellaneous rubbish; or a disorderly accumulation of things lying about (Fig. 66), giving a general appearance of untidiness and lack of attention to good HOUSEKEEPING (*q.v.*). See also PERIMETER AREA.

In mammals, the group of young produced at birth (Fig. 33), the number varying considerably, e.g. in rats 5–10 young, and mice 4–16.

Liver baiting. (*proc*) The laying of fresh liver, in centimetre cubes, on small pieces of paper or aluminium foil, overnight, to indicate the extent of infestations of Pharaoh's ant, (*Monomorium pharaonis*). This technique frequently identifies the approximate positions of nests. Also the use of minced liver admixed with insecticide (e.g. chlordecone) for ant control.

Local authority. (*name*) A local body of elected officers responsible for applying national legislation at local level, together with local bye-laws. An essential activity is to maintain required standards of public health (see PUBLIC HEALTH INSPECTOR). Many authorities employ their own rodent operators; others use contractors to carry out pest control work.

Longevity. (*zoo, chem*) The length of life, or life expectancy of an animal. The duration of activity of an insecticide or rodenticide in use in the field.

Long-headed flour beetle. (*ent*) See LATHETICUS ORYZAE.

Long-tailed field mouse. (*zoo*) See APODEMUS SYLVATICUS.

Loxosceles. (*zoo*) Arachnida: Araneae. A genus of Brown spiders, common in many areas of the United States, the bite causing a necrotic skin lesion—a slow-healing wound leading to the formation of an abscess, usually leaving a sunken scar. An intense pain follows some hours after the bite, which is rarely fatal. Brown spiders live both indoors and out hiding in crevices in the dark, in crawl spaces, basements, garages, and concrete block buildings. Other locations include grain bins and boxes containing transformers and electrical switch gear.

All species of *Loxosceles* should be regarded as dangerous to man.

L. reclusa, the Brown recluse spider, (10 mm long) is yellow-brown with a dark brown-black area around the eyes, extending backward as a thin narrow line over the abdomen—the whole being 'fiddle'-shaped. The legs are exceptionally long and slim in relation to body size. Eggs are laid in off-white silken sacs, the young, usually about 40, taking about 1 month to hatch and 7–12 months to mature.

L. unicolor, despite the name Brown spider, has the back pale to bright yellow, head orange, and legs yellow with orange to red ends. It lacks the fiddle-shaped marking.

Lucilia sericata. (*ent*) See GREEN BOTTLE FLIES.

Lyctus brunneus. (*ent*) Coleoptera: Lyctidae. Powder post beetle. An occasional pest of timber in buildings in the U.K. (in less than 1% of buildings surveyed) occurring mainly in new wood block floors and furniture. Infestations occur outdoors in partially seasoned hardwoods, notably in timber yards. Only the sapwood of hardwoods is attacked.

The adult (4 mm long) is reddish-dark brown, and flies actively. Eggs are laid deep in the pores of wide-pored hardwoods by means of a slender ovipositor, and hatch in 2–3 weeks. Larvae (5–6 mm long) have a conspicuous brown spiracle on each side of the eighth abdominal segment which is three or more times larger than those on segments 1–7. Larvae mature in about 1 year. The wood is completely powdered with little signs of individual tunnels. The boredust is flour-like without distinct pellets (cf. ANOBIUM PUNCTATUM and XESTOBIUM RUFOVILLOSUM).

The larvae feed on the contents of the cells of the timber; starches and sugars are digested but not cellulose. Pupation in the timber is complete in 2–4 weeks; adults emerge in May–September (U.K.); exit holes are about 1 mm diam. Two generations per year may occur in timbers with high starch content.

M

MAC value. (*tox*) The maximum allowable concentration of a substance that can be inhaled for 8 hours a day for many years without ill effect. The values for different pesticides vary in different countries.

Madeira cockroach. (*ent*) See LEUCOPHAEA MADERAE.

Maggot. (*ent*) A non-technical term: strictly the larva of a fly (Diptera, Fig. 61), but also the immature stage of other insects in which the larva is legless and progresses by a wriggling movement. Cf. CATERPILLAR and GRUB.

Maintenance. (*bldg*) The process of keeping in good repair, essential for buildings used for food storage or processing. Responsibility should be allocated to a maintenance engineer with an understanding of pests and preventive pest control. To this end, attention becomes necessary: to the building structure (rodent proofing, insulation, weather proofing and ventilation); to drainage (its cleaning and proper functioning); to waste (its accumulation and daily removal); to machinery, elevators and conveyors (their proper functioning, prevention of spillage, ease of cleaning and dust control); to all services (such as plumbing and electricity); and to the building surrounds and their maintenance in a tidy and clean condition. See PERIMETER AREA.

Maize weevil. (*ent*) See SITOPHILUS ZEAMAIS.

Malathion. (*chem*) An organophosphorus compound with a broad spectrum of insecticidal activity introduced by American Cyanamid in 1950 ('Mercaptothion' in West Germany and South Africa, and 'Carbofos' in the U.S.S.R.). A colourless to light amber liquid of limited solubility in water but miscible with most organic solvents. Corrosive to iron; lined containers must be used. Degradation is rapid on alkaline surfaces, otherwise it has moderate persistence.

Malathion has good acaricidal properties. Its principal use is in the control of stored food pests, as a GRAIN PROTECTANT (*q.v.*) and for the control of bedbugs, cockroaches, fleas, flies, mosquitoes, ticks and a number of household insects.

The mammalian toxicity is low: oral LD_{50} (rat) 1500 mg/kg. It is not readily absorbed through the skin. Formulations available include emulsion concentrates, wettable powders and dusts. Deodorised grades of malathion have now replaced earlier agricultural grades which had an objectionable odour.

Mallophaga. (*ent*) The Order of insects containing the biting and chewing lice; external parasites mainly of birds, occasionally of mammals. Small, flattened, wingless insects, the head always wider than the thorax, the mouthparts modified for chewing skin, hair or feathers of birds. The entire life cycle is spent on the body of the host. The Mallophaga contains no known parasites of man; the insects are unable to survive when removed from the host. Control measures are rarely required. The most usual biting louse found indoors is the Dog biting louse, *Trichodectes canis*.

Mammalia. (*zoo*) A class of vertebrates in which the females have milk-producing glands on the ventral surface. Additional features which separate mammals from other vertebrates include the presence of hair, a diaphragm used in respiration and three bones in the middle ear. The most important pest mammals are rodents.

Mammalian toxicity. (*tox*) A measure of the harmful effect of a substance on a mammal, normally quoted as the LD_{50} (*q.v.*) with reference to the species used, usually the rat. But see also MINIMUM LETHAL DOSE and LETHAL DOSE.

Mandible. (*ent*) Of insects, one of the pair of mouthparts which does most of the work in cutting and crushing food (e.g. cockroaches). In some insects they are wanting (as in most flies), or much reduced (as in most adult Lepidoptera).

Marlate. (*p. prod*) See METHOXYCHLOR.

Martin fly. (*ent*) See CRATAERINA PALLIDA.

Mating. (*zoo*) The process of pairing (copulation) for breeding purposes. A number of factors bring the sexes together: behavioural, (e.g. courtship display), physical (e.g. coloration) and chemical (odorous attractants). In some animals mating is seasonal (some birds), in others (insects) it may be frequent throughout adult life.

Maxilla. (*ent*) Of insects, one of the pair of mouthparts lying behind the mandibles, usually with a segmented palp (maxillary palp). The maxillae assist the mandibles in holding and masticating food.

May bug. (*ent*) See CHAFERS.

MCA-600. (*chem*) See MOBAM.

Mealworm. (*ent*) A general term, usually meaning the larva of the Yellow mealworm beetle (see TENEBRIO MOLITOR), the Dark mealworm beetle (see TENEBRIO OBSCURUS), but occasionally the Lesser mealworm beetle (see ALPHITOBIUS DIAPERINUS).

Medical Officer of Health. (*name*) An officer of Local Government responsible for all aspects of health (and pest control) in his area, operating with the help of PUBLIC HEALTH INSPECTORS (*q.v.*).

Mediterranean flour moth. (*ent*) See EPHESTIA KÜHNIELLA.

Melting point. (*phy*) The temperature at which a solid changes to a liquid (e.g. 2·8°C for malathion, 175°C for dieldrin).

Mercaptothion. (*p. prod*) See MALATHION.

Merchant grain beetle. (*ent*) See ORYZAEPHILUS MERCATOR.

Mesothorax. (*ent*) The second, middle segment of the thorax of an insect, immediately behind the prothorax and bearing the second pair of legs and in alate forms the first pair of wings (or elytra).

Mestranol. (*chem*) A steroid compound of use in human birth control and of suggested value as a chemosterilant for rodents; in baits it requires repeated feeding to cause permanent sterility; field tests show poor bait acceptance. No chemosterilant has yet shown practical value for commensal rodent control, but see ORNITROL.

Metabolism. (*zoo*) The chemical processes which occur within an organism. Pesticides interfere with some vital metabolic processes bringing about death.

Metadelphine. (*p. prod*) See DIETHYL TOLUAMIDE.

Metamorphosis. (*zoo*) A characteristic of most insects; a change of form during development from the egg to the adult. Two types are recognised: 1) 'complete': egg, larva, pupa, adult and 2) 'incomplete': lacking a pupal stage, the immature insects referred to as nymphs.

Metathorax. (*ent*) The hind most of the three thoracic segments of an insect, immediately behind the mesothorax, bearing the third pair of legs, and in alate forms the second pair of wings (halteres in flies).

Methoxychlor. (*chem*) An organochlorine insecticide introduced in 1945 by Geigy under the trade name 'Marlate'. In technical form, a grey flaky powder, resistant to heat and oxidation. It has a range of insecticidal activity similar to DDT but, in contrast, shows little or no accumulation in animal body fat, or excretion in milk. As a result, methoxychlor has been recommended as a replacement for DDT for fly control in dairy barns. Other uses include the control of carpet beetles, clothes moths, fleas and insect pests of poultry houses.

The acute oral LD_{50} (rat) is very low: 6000 mg/kg. Formulations available include wettable powders and emulsion concentrates.

Methyl bromide. (*chem*) Bromomethane. A colourless liquid changing to a colourless gas at 4·5°C, used as a fumigant. Its high insecticidal properties were first reported in 1932; it is stable, non-corrosive and non-flammable but degrades natural rubber.

It is supplied in pressure cylinders as a liquid; also in metal cans and glass ampoules. The gas is three times heavier than air.

Methyl bromide is used for space and commodity fumigation: for the control of insects and mites in food stores, for the treatment of mills, ships and for the fumigation of soils against nematodes, fungi and weed seeds. Methyl bromide penetrates quickly and deeply into most commodities and airs off rapidly after treatment. Foods with a high oil content, e.g. nuts, sago flour, should be treated with care since absorbed bromide produces taint.

The gas is highly toxic to man, not detectable by smell at normal fumigation concentrations; a warning gas, CHLOROPICRIN (*q.v.*) is therefore usually incorporated at 2%. The safe upper limit, above which respirators must be worn is 17 ppm. Unprotected exposure to 100–400 ppm for only a few hours may cause severe illness or death. The minimum concentration of methyl bromide that can be detected by a HALIDE LAMP (*q.v.*) is 10 ppm. prolonged contact with the skin, e.g. through contaminated clothing or footwear, can cause severe blistering. Use of methyl bromide is restricted in many countries to trained and experienced personnel.

Methyl formate. (*chem*) A colourless, highly inflammable liquid of pleasant odour used as an insecticidal fumigant for the individual treatment of packages of dried fruits and nuts. It is applied as a liquid and allowed to vapourise within the container; maximum insecticidal effect against some insects is obtained by admixture with 40% carbon dioxide. Dilution with CO_2 also reduces the fire hazard.

The vapour is toxic to man: the safe concentration in air for indefinite exposure is thought to be 1500 ppm. The maximum tolerable concentration for a 1-hour exposure is 4000 ppm.

Mezium. (*ent*) Coleoptera: Ptinidae. A genus of spider beetles, containing *M, affine* and *M. americana*. Adults (2–3 mm long), with the thorax and head clothed with golden scales and hairs; the elytra strongly convex, shining and dark red-brown. Occur in homes, warehouses and granaries. These beetles feed on animal and vegetable debris.

MGK repellents. (*p. prod*) A group of unrelated compounds marketed by McLaughlin Gormley King Co. (and Phillips Petroleum Co.) under various trade names as insect repellents. They include:

MGK Repellent 11: Introduced in 1949 as a repellent against cockroaches and mosquitoes. Also in combination with pyrethroids mainly for the protection of dairy cattle against flies. The technical product is a pale yellow liquid with a fruity odour, insoluble in water, miscible with some organic solvents. The acute oral LD_{50} (rat) is 2500 mg/kg.

MGK Repellent 326: First described in 1953; used as a repellent against houseflies. The technical product is an amber liquid with a mild aromatic odour. Practically insoluble in water, miscible with alcohols and kerosene. Unstable to sunlight, non-corrosive, hydrolysed by alkalis. The acute oral LD_{50} (rat) is 6000 mg/kg. Formulated as oil sprays and emulsions up to 5% active ingredient.

MGK Repellent 874: Also described in 1953; chiefly used as a cockroach repellent. A light amber liquid with a mild sulphurous odour. Slightly soluble in water, miscible with most organic solvents. Stable, non-corrosive, compatible with many insecticides. The acute oral LD_{50} (rat) is 8500 mg/kg. Applied in oil or emulsion (1–5%) at 1000 mg active ingredient/m^2.

MGK 264. (*p. prod*) A synergist for pyrethroids and carbamates, first described in 1949 and subsequently marketed by McLaughlin Gormley King Co. A viscous liquid, practically insoluble in water, but soluble in most organic solvents; stable, non-corrosive and compatible with most pesticides. Acute oral LD_{50} (rat) is 2800 mg/kg. Usually incorporated at 0·5–2% in aerosols, oil sprays, emulsions and dusts; not as active as PIPERONYL BUTOXIDE (*q.v.*) in synergising pyrethrins.

Mice. (*zoo*) See MUS MUSCULUS and APODEMUS SYLVATICUS.

Micro-drop technique. (*proc*) See TEST METHODS.

Microencapsulation. (*proc*) The technique of forming a thin wall of material around particles of a pesticide to enhance its performance, e.g. of a rodenticide to improve palatability of baits and delay speed of absorption in the gut, or of an insecticide to reduce dermal hazard or resistance to degradation in the environment (Fig. 57). Application of this process to pest control has so far been largely experimental.

Microsol generator. (*equip*) See FOGGING MACHINES.

Migration. (*zoo*) The mass movement of individuals of a species due to environmental factors, or, as an inherent feature of behaviour. The first has been observed (e.g. in cockroaches) in response to population pressure and flooding where movement is multidirectional. The wandering behaviour of large numbers of moth or fly larvae in search of pupation sites is sometimes described as migration. The second is characteristic of some butterflies and birds where the direction of movement is well-defined.

Millipedes. (*zoo*) See DIPLOPODA.

Mill moth. (*ent*) See EPHESTIA KÜHNIELLA.

Minimum Lethal Dose. (*tox*) MLD. The lowest of a series of graded doses which kills at least one individual in a test group.

Mirex. (*chem*) A chlorinated ketone introduced as an insecticide in 1959 by Allied Chemical Corporation. Related to CHLORDECONE (*q.v.*) but chemically unlike other chlorinated hydrocarbon insecticides.

The technical form is a white solid of very low volatility. It is formulated in baits (0·075–0·15% active ingredient) as a stomach poison for the control of ants; principally, Argentine ant, Imported fire ant, Western harvester ant and Leaf cutter ants. Also used to prevent attack by fabric-feeding insects.

The acute oral LD_{50} (rat) is 600 mg/kg with a low order of toxicity to birds and fish.

Misting. (*proc*) The application of an insecticidal spray to a surface in such a way as not to make it obviously wet (Fig. 58). It involves the use of a fine nozzle to produce minute droplets. Especially important on textiles and other surfaces which may be susceptible to wetting and perhaps damage by heavier spraying. See U.L.V.

Mites. (*zoo*) See ACARI.

Mobam. (*chem*) MCA-600. A carbamate insecticide of low toxicity, similar in structure to carbaryl, a product of Mobil Chemical Co., released in recent years for experimental evaluation. A promising compound for cockroach control.

Moisture content. (*phy*) The weight of water in a substance (e.g. grain) expressed as a percentage of the dry weight. Important in dry climates in relation, for example, to the use of ALUMINIUM PHOSPHIDE (*q.v.*) as a fumigant, or the performance of CYMAG (*q.v.*) in the treatment of rat burrows.

Mole cricket. (*ent*) See CRICKETS.

Monomorium pharaonis. (*ent*) Hymenoptera: Formicidae. Pharaoh's ant. A small tropical species (2 mm) which has taken advantage of heated premises provided by man, e.g. hospitals, restaurants, and institutional buildings, to establish large infestations in the hollow spaces, often inaccessible, behind stoves, boilers and hot water pipes. Foraging occurs over large areas, trails following the course of hot water pipes and mortar jointing of tiled areas. Protein foods are preferred; water is essential. In warm climates, nests may be made outdoors.

This ant is yellow-brown with the tip of the abdomen darker; there are two segments to the pedicel; development time from egg to adult is about 7 weeks. There may be a number of queens in one colony. A pest often difficult to eradicate; insecticidal baits are usually the most effective. See LIVER BAITING.

Morphology. (*zoo*) The study of the external form of animals. Knowledge of the morphological characters of pests is necessary for correct identification. Also sometimes known as external anatomy.

Mortality. (*tox*) Death, kill, the end result of properly applied pesticides.

Mosquitoes. (*ent*) See CULICIDAE.

Moth balls. (*chem*) See NAPHTHALENE.

Moths. (*ent*) See LEPIDOPTERA.

Moulting. (*zoo*) In insects, loss of the cuticle during metamorphosis, now considered to consist of apolysis (the separation of the outer cuticle from the epidermal cells) and ecdysis (the act of shedding the old cuticle); occasioned by the rigidity of the integument which is ill-adapted for accommodating the increase in size of the insect. An event which occurs between successive larval and nymphal stages. The number of moults differs in pest insects and is often variable within the same species.

In birds and mammals, the process of shedding feathers and hair from parts of the body, which may become important contaminants of food.

Mud tube. (*ent*) See SHELTER TUBE.

Murine typhus. (*dis*) An infection of man, rodents (usually Brown and Ship rats) and other small mammals, spread by lice, fleas and mites. The causative pathogen is *Rickettsia typhi*; the chief carrier is the Oriental rat flea *Xenopsylla cheopis*. This disease has certain similarities to plague but outbreaks are usually much less extensive.

Musca autumnalis. (*ent*) Diptera: Muscidae. The Autumn fly or Face fly. A troublesome cluster fly in the autumn when it enters houses and again in spring when it leaves. The adult female is similar in appearance to *Musca domestica* but is a little larger (6 mm long). The abdomen of the male is orange with a dark stripe down the middle.

Eggs are laid on cow dung in which the larvae develop. The adults feed on the sweat of cattle in summer and on lesions produced by other biting flies. See CLUSTER FLIES.

Musca domestica. (*ent*) Diptera: Muscidae. The House fly. A world-wide pest of homes, shops, factories, catering establishments and rubbish tips, the adult attracted to, and breeding in, decaying animal and vegetable waste. A major health risk: a carrier of a large number of disease organisms, (e.g. the bacteria causing food poisoning and infantile diarrhoea). The characteristic behaviour of short flights followed by settling and exploration of surfaces (cf. FANNIA) helps to spread disease. Active by day, resting at night, preferring projecting edges high up in rooms as alighting surfaces.

The adult (6 mm long) has a grey thorax with four narrow black stripes on the upper surface (Fig. 60). Distinguished from some other indoor flies by the fourth vein being sharply bent forward to meet the wing edge just behind the vein in front.

Eggs are banana-shaped, laid in moist decaying vegetable matter, indoors and out, rarely in domestic dustbins, and under favourable conditions eggs hatch in 8 hours. The larvae, 'maggots' (Fig. 61), are white, legless, (12 mm long) and migrate to dry locations to pupate (Fig. 62). The adult lives for 1–3 months and begins to lay eggs 2 days after emergence. Minimum development time from egg to adult can be as short as 8–9 days. Houseflies over-winter in the larval, pupal and adult stages.

Mus musculus. (zoo) House mouse (Fig. 63). A world-wide pest of buildings; distinguished from rats by smaller size and from young rats by the larger ears of the mouse, much longer tail, and smaller head and feet relative to the body. The House mouse is grey above, light grey below; normally lives indoors, preferring dry nesting locations; usually ground living but can climb. Movement in search for food is very limited, nests often occurring within the foodstuff itself, and frequently made from shredded paper. Food from a number of sources is taken during a night's foraging. Cereals are preferred; grains are kibbled (see FEEDING BEHAVIOUR).

The gestation period is about 3 weeks, the size of litters varies considerably (4–16); the number of litters in a year averages 7–8, the period from birth to sexual maturity is 8–10 weeks. Longevity is between 6–12 months. The maximum reproductive potential of a pair and their young in a year is about 2000. Much more prolific than rats.

Resistance to warfarin is widespread in mice in the U.K. at the present time. *Mus musculus* has therefore become a primary pest in all types of buildings. See RODENT CONTROL.

Myriapoda. (zoo) An obsolete name for a Class of Arthropods containing centipedes and millipedes, now assigned to separate Classes. See CHILOPODA and DIPLOPODA.

Naled. (*chem*) An organophosphorus compound, introduced in 1956 by Chevron Chem. Corp. under the trade name 'Dibrom'. One of the less toxic organophosphorus insecticides, of short residual life, with little application except as a space spray for the control of flying insects (e.g. in mushroom houses at 1 g/10 m³). Also for outdoor mosquito control.

The technical material is a yellow liquid with a slight but pungent odour readily hydrolised in water. In the presence of metals and reducing agents, Naled rapidly loses bromine and reverts to DICHLORVOS (*q.v.*).

The acute oral LD_{50} (rat) is about 350 mg/kg. Formulations include emulsion concentrate and dust.

Nankor. (*p. prod*) See FENCHLORPHOS.

Naphthalene. (*chem*) Moth balls. Colourless crystalline solid, long used as a household fumigant for protecting furs and woollens against moth, but of low potency, having doubtful value as a moth repellent with limited insecticidal properties. Insoluble in water, soluble in most organic solvents. Inflammable but safe at ordinary temperatures.

Toxicity to mammals is low: acute oral LD_{50} (rat) 2200 mg/kg; the fatal dose to man is 2–3 g. High concentrations of the vapour are damaging to the eyes and crystals are irritant in contact with the skin and mucous membranes. These symptoms give naphthalene self-warning properties.

Formulations available include flake naphthalene, and compressed moulded 'cakes', granules and moth balls.

Nasute. (*ent*) See SOLDIER TERMITE.

National Pest Control Association. (*name*) N.P.C.A. The trade association of the structural pest control industry of the United States, formed in 1933 as the National Association of Exterminators and Fumigators, renamed in 1937. Membership includes 1300 active members on the U.S. mainland and 170 firms elsewhere in the world.

Staff monitor legislative, regulatory, technical and management developments among federal and state governments, universities, research agencies, and the business community, thus keeping the industry advised through letters, releases and reports. A board of directors develops Association policies and committees develop programmes and good practice statements on industry work. A self-instructional training programme for service personnel is currently under development. The aim is to serve those needs of the industry that cannot be met on a practical basis through the effort of individual member companies.

Secretary: 250 West Jersey Street, Elizabeth, New Jersey 07207, U.S.A.

National Poisons Information Centre. (*name*) A medical department with whom information on pesticide formulations is deposited, in the event of it being required in cases of accidental poisoning. Such centres in the U.K. are:

Poisons Reference Service,
New Cross Hospital,
Avonley Road,
London, S.E.14.
Telephone 01–407 7600 (Poisons).

The Poisons Information Centre,
Cardiff Royal Infirmary,
Newport Road,
Cardiff.
Telephone Cardiff 33101.

Scottish Poisons Information
Bureau, Royal Infirmary,
Edinburgh 3.
Telephone 031–229 2477.

The Poisons Information Centre,
Royal Victoria Hospital,
Belfast.
Telphone Belfast 30503.

National Research Development Corporation. (*name*) N.R.D.C. Promotes the adoption by industry of new products and processes invented in U.K. government laboratories, universities and elsewhere, advancing money where necessary to bring them to a commercially viable stage. Wide ranging in its activities in the entire field of science and technology. It speeds up technological advance by investing money with industrial firms for the development of their own inventions and projects.

N.R.D.C. does not itself manufacture or trade, nor does it have its own research or development facilities. It does not receive annual grants but is financed by the Secretary of State for Trade and Industry with government loans. A minor part of the activities of N.R.D.C. includes the promotion of new pesticides and pest control techniques. Original patent holders in the U.K. for insecticidal lacquers.

Necrobia rufipes. (*ent*) Coleoptera: Cleridae. The Copra beetle. A pest of warehouses and imported cargoes of copra (dried coconut), dried meats and fishmeal. Prevalent in ham and bacon curing premises and dog biscuit factories.

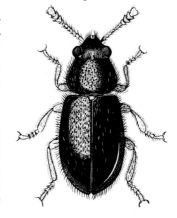

The adult (5–7 mm long) has a distinctive green-blue body with reddish legs. Adults run rapidly but rarely fly. The eggs (1 mm long), smooth, shining and translucent hatch in 4–5 days. The larva has a dark brown head and thorax, the remainder pale with bluish marks. The pupa is contained in a paper-like cocoon. The minimum period of development from egg to adult is about 30 days, extending to many months under less favourable conditions. Larvae over-winter in un-heated premises.

Neocid. (*p. prod*) See DDT.

Neopynamin. (*p. prod*) See Tetramethrin.

Nervous system. (*zoo*) The route through which external stimuli (e.g. light, noise, odours) perceived by the sense organs, cause an organism to avoid unfavourable situations in its environment and react positively towards favourable conditions. In insects, it is composed of three inter-related systems; the central nervous system (the brain, ventral nerve cord and ganglia), the peripheral nervous system (sensory structures and their nerve fibres) and the sympathetic nervous system (having both nervous and hormonal activity).

Nest. (*zoo*) A structure, often of collected material, providing shelter, over a period of time for egg-laying or for the rearing and nursing of young; applicable to birds and rodents but distinct from 'harbourage', a tem-porary shelter used as a retreat or resting site (as in cockroaches).

Nexion. (*p. prod*) See Bromophos.

Night inspection. (*proc*) The survey of a building at night, notably for cockroach infestation. Cockroaches remain hidden in crevices during the day; the same premises visited at night give a much more reliable impression of the size and distribution of an infestation, because of greater cockroach activity (see Circadian rhythm). Night inspection also reveals the extent of residual infestations following insecticide application.

Niptus hololeucus. (*ent*) Coleoptera: Ptinidae. The Golden spider beetle. A pest of temperate regions, the larvae feeding on vegetable and animal debris in warehouses, cellars and ill-kept stores, rarely an important pest of the stored food itself. The adult can be troublesome in homes by biting holes in clothing, carpets and bedding, moving actively at temperatures as low as 5°C. High temperatures are not tolerated.

The adult (3–4 mm long) has long golden-yellow silky hairs covering fused elytra. The head is concealed under a rounded thorax. Develop-ment from egg to adult takes about 4 months (at 20°C). Adults live 7–10 months.

Nitidulidae. (*ent*) The family of the Coleoptera containing the Dried fruit beetles. See Carpophilus.

Nitrochloroform. (*chem*) See Chloropicrin.

Non-stop aerosol. (*equip*) An aerosol pack from which the entire contents are discharged when the valve is released. Usually containing pyrethrins or dichlorvos. Used for the space treatment of storage rooms. Special precautions apply; masks must be worn.

Norbormide. (*chem*) A carboximide compound introduced as an acute rodenticide in 1965 by McNeil Labs. Inc. under the trade names 'Raticate' and 'Shoxin'. Specific to *Rattus* spp.

An off-white crystalline powder with low solubility in water, acting as a vasoconstrictor, inhibiting the flow of blood to the terminal capillaries and interfering with heart function. The acute oral LD_{50} (rat) is 12 mg/kg for *R. norvegicus*, and 50 mg/kg for *R. rattus*. Poisoning symptoms appear quickly. Mice survive 1000 mg/kg.

When this compound was introduced, the promise of a specific rodenticide was received with enthusiasm. However, too rapid action and bait shyness, severely limit the value of norbormide with the result that current use is minimal. Formulations include concentrates and small sachets of ready-to-use bait containing 0·5–1·0% active ingredient.

Northern rat flea. (*ent*) See NOSOPSYLLUS FASCIATUS.

Norway rat. (*zoo*) See RATTUS NORVEGICUS.

Nosopsyllus. (*ent*) Siphonaptera: Ceratophyllidae. A genus of rodent fleas. *N. fasciatus*, the European or Northern rat flea, is primarily found on *Rattus norvegicus*, but sometimes occurs on mice and voles. *N. londiniensis* is mainly associated with *Mus musculus* and *Rattus rattus*. Both species are cosmopolitan, particularly common in sea ports.

Notification of pesticides. (*leg*) See APPROVAL OF PESTICIDES.

Nuvan. (*p. prod*) See DICHLORVOS.

Nuvanol N. (*p. prod*) See IODOFENPHOS.

Nymph. (*zoo*) Immature stage of an insect with an incomplete metamorphosis (no pupal stage), hatching from an egg and changing by a series of moults directly into an adult. Characterised by having a form and habitat very similar to the adult (Fig. 64). Wings are present as external buds (cf. LARVA) but are non-functional until the adult stage is reached. Also the immature stages of mites (excluding the first stage larva). Pest insects having nymphal stages include cockroaches, bedbugs, booklice, silverfish and termites.

O

Octachlor. (*p. prod*) See CHLORDANE.

Octalene. (*p. prod*) See ALDRIN.

ODCB. (*chem*) See ORTHODICHLOROBENZENE.

Odourless kerosene. (*chem*) A petroleum distillate with a narrow boiling range, from which the more odorous fractions have been removed. A useful solvent and carrier for many pesticides; of moderate volatility. Reacts with many types of surface encountered in buildings, e.g. polystyrene and vinyl floor coverings. If staining occurs, marks disappear on drying.

Oil base for insecticides. (*chem*) A variety of oils is used for this purpose, depending on type of application; smell, flammability and solvent characteristics are of importance. Oils used for timber preservation are preferably of low volatility to allow maximum insecticide/fungicide penetration. Boiling range and aromatic content determine the potential risk of carcinogenicity. See also ODOURLESS KEROSENE and OIL SPRAY.

Oil spray. (*chem*) An insecticidal formulation in which the active ingredient is dissolved in, or is miscible with oil (e.g. odourless kerosene) as the DILUENT (*q.v.*). The solubility of incorporated solids may sometimes be increased by the addition of co-solvents. Most oil sprays have the advantages of being ready to use with little risk of staining. They have many disadvantages; relatively high cost, flammability, speedy penetration of porous surfaces and greater bulk to transport. Cf. EMULSION CONCENTRATE and WETTABLE POWDER.

Old house borer. (*ent*) See HYLOTRUPES BAJULUS.

Oötheca. (*ent*) Egg case. A sac-like structure with various ornamentations, containing the eggs of cockroaches (and mantids), protecting the developing young. The 'keel' on the dorsal surface bears tiny openings which allow the embryos to breathe.

Size, colour and shape of the oötheca varies in different species; also the number of eggs contained. Oöthecae of the German, and Brown-banded cockroaches are 'segmented' indicating the position of the eggs within. Those of the Oriental cockroach and *Periplaneta* species are not so marked.

The material of the oötheca is produced by certain glands (colleterial) associated with the female reproductive system. Some cockroaches retain the oötheca attached to the female until the eggs hatch (e.g.

Blattella germanica), others deposit and attach it to a suitable surface soon after formation (e.g. *Blatta orientalis* and *Periplaneta* spp.). A few pest species (e.g. *Leucophaea maderae*) retain the oötheca within the body while the eggs develop.

OP insecticide. (*chem*) See ORGANOPHOSPHORUS INSECTICIDE.

Operator. (*name*) See SERVICEMAN.

Oral. (*tox*) By the mouth. A portal of entry of pesticides into the body.

Oral dosing. (*proc*) See TEST METHODS.

Organic thiocyanates. (*chem*) See LETHANE 384.

Organochlorine insecticide. (*chem*) A hydrocarbon insecticide incorporating chlorine. Compounds of this type act on the nervous system of insects causing loss of co-ordination. Similar effects are produced in man and other animals. Most organochlorines accumulate in fatty tissues and are highly persistent in the soil. Examples of organochlorine insecticides are DDT and dieldrin (which have made a major contribution to insect control but are now banned from certain uses in many countries), chlordane, and lindane. The organochlorines were the first group of synthetic insecticides to be developed during the Second World War, but are now widely criticised for undue persistence in the environment and accumulation in biological food chains.

Organofluorine compounds. (*chem*) See FLUORACETAMIDE and SODIUM MONOFLUOROACETATE.

Organophosphorus insecticide. (*chem*) An insecticide derived from phosphoric acid. A large number of compounds belong to this group (the organo-phosphates), varying widely in insecticidal activity and toxicity to man. They act on the nervous system by inhibiting the enzyme, cholinesterase, thus interfering with the normal mechanism of nerve impulse transmission. They are not as persistent as the organochlorines but are as toxic when absorbed through the skin as when taken orally. More dangerous as acute poisons than as chronic poisons.

Examples of the group which have made a valuable contribution to industrial and domestic pest control are malathion, diazinon, fenitrothion and dichlorvos, the latter with the useful property of high volatility.

Oriental cockroach. (*ent*) See BLATTA ORIENTALIS.

Oriental rat flea. (*ent*) See XENOPSYLLA CHEOPIS.

Ornitrol. (*p. prod*) The trade name of G. D. Searle & Co. for an avian sterilant used for population control of feral pigeons; an analogue of cholesterol blocking the production of cholesterol in the body. It is fed as a whole grain bait containing 0·1% active ingredient for 10 days and inhibits reproduction for about 6 months.

The technical product is an odourless, white crystalline solid; moderately soluble in water, but insoluble in oils. Doses of 10 mg/kg per day fed to rats for 4 weeks produced little evidence of any harmful action.

Orthoboric acid. (*chem*) See BORIC ACID.

Orthodichlorobenzene. (*chem*) ODCB. A liquid with a pungent characteristic odour, insoluble in water but miscible with many organic solvents, once used as a co-solvent for insecticides in timber preservatives, the ODCB providing fumigant action and emetic properties in accidental ingestion. Maximum allowable concentration for prolonged exposure 50 ppm.

Orthoptera. (*ent*) The Order of insects containing the grasshoppers, locusts and crickets, capable of jumping. The latter are occasionally pests in buildings. See CRICKETS.

Oryzaephilus. (*ent*) Coleoptera: Silvanidae. A genus containing two pest species; general feeders on cereal debris, rarely attacking whole grain until damaged by other insects. Body (2–3 mm long) narrow and flattened, light red to black. Thorax with distinct saw-toothed edge (6 teeth) and three ridges on the upper surface. Active beetles, running rapidly, the eggs laid loosely in the food. Minimum period of development from egg to adult 20 days (35°C), extending to 3–4 months (20°C). Little development or egg laying occurs below 18°C.

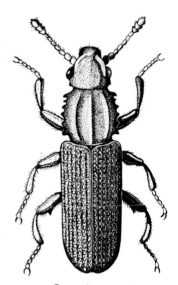

O. surinamensis

Oryzaephilus surinamensis: the Saw-toothed grain beetle. The major pest of grain stored on farms in the U.K. Encouraged by the retention of grain at high moisture content and inadequate cooling after artificial drying. *O. surinamensis* is distinguished from the following by the eyes being not so close to the rear margin of the head.

Oryzaephilus mercator: the Merchant grain beetle. A minor pest of imported cereals and nuts, not over-wintering in the U.K. The eyes closer to the hind margin of the head than in the Saw-toothed grain beetle.

Otiorrhynchus. (*ent*) Coleoptera: Curculionidae. A genus of Garden weevils, including the common species, *Otiorrhynchus rugostriatus, O. sulcatus* and *O. singularis.* Occasionally pests in gardens where strawberries and greenhouse plants may be attacked. Common casual intruders of homes, especially in the autumn. The adults are 6–8 mm long

O. rugostriatus

and have a short, thick rostrum (snout) with elbowed antennae. The larvae are white, fleshy and legless, feeding inside plant tissues. Garden weevils do not breed indoors and no treatment is necessary against adults which may occasionally wander in.

Oviposition. (*ent*) The process of egg-laying, including, in some insects, the selection of a suitable egg-laying site, the forming of a cavity in which to deposit the egg(s), the extrusion of the egg, and securing to surfaces. Some pest insects deposit their eggs loosely in foodstuffs. In cockroaches, oviposition is the passage of eggs into the OÖTHECA (*q.v.*).

Ovum. (*zoo*) pl. Ova. Strictly an unfertilised egg formed by oögenesis in the ovary. Often used to describe the fertilised eggs of some insects (Lepidoptera). See EGG.

Owl midges. (*ent*) See PSYCHODIDAE.

Packaging materials. (*manuf*) Materials in which goods are consigned or presented for display, varying with the merchandise, but consisting variously of packing straw, lightweight polystyrene, corrugated cardboard, timber crates, metal and plastic casings, and laundry baskets.

Of importance in pest control in providing harbourage for rodents and insects and their possible introduction into all types of premises including the home. Industries handling products in corrugated cardboard are particularly at risk: corrugations provide a multiplicity of harbourages for cockroaches and booklice; and pupation sites, especially for pest moths (Fig. 49). Beverage crates provide favourable harbourage for cockroaches and their transfer between bottling plants, breweries, hotels, cafés, restaurants and public halls.

Returnable crates should always be heat sterilised or treated with insecticide before re-use. Packaging materials should never be allowed on production floors (Fig. 65), where they may introduce infestation and consequent contamination of manufacturing equipment and products being produced.

Pain. (*zoo*) See ANIMALS (CRUEL POISONS) ACT, 1962.

Palatability. (*zoo*) The appreciation of an animal, by way of chemo-receptors on the tongue and palate, for a food flavour and texture. An important consideration in the formulation of rodent baits to improve 'take'. See also ADDITIVE.

Pallet. (*manuf*) A double-sided timber raft usually not more than 1·5 m square, used for ease of handling commodities by fork lift truck or other equipment; a means of keeping commodities off the ground during storage (Fig. 78). Pallets should never be placed so close in storage as to prevent inspection of goods at intervals along a stack, or so close to walls as to prevent access between the wall and the stack (Fig. 51). The dead space beneath pallets provides favourable harbourage and runs for rodents; these are ideal locations for siting rodenticidal baits. See also STACKING.

PAM or **2 PAM.** (*p. prod*) See PRALIDOXIME CHLORIDE.

Pantry pests. (*ent*) A miscellaneous group of insects, introduced into the larder, pantry or food cupboards of homes and catering establishments by purchase of infested products (sometimes from local shops) and often introduced in packaging materials. Infestations develop in products left on shelves for extended periods, the insects cross-contaminating other products, some species capable of gaining entry even though the packages may be unopened. The group includes various beetles, moths and occasionally mites if the foodstuff is damp.

Control is best achieved by taking all foods out of the store, inspecting them thoroughly, and getting agreement with the housewife, or proprietor, to discard all that are infested. The empty larder, cupboards and shelves should then be sprayed (with e.g. diazinon, malathion or lindane) and allowed to dry and the space ventilated. Products free from attack should then be replaced, lining the shelves first with paper to prevent contamination. SLOW RELEASE STRIPS (*q.v.*) are not recommended where food is exposed.

Paradichlorobenzene. (*chem*) An insecticidal fumigant which first came into use in 1913. Used primarily to control clothes moths, carpet beetles and dermestids, and to prevent fly-breeding in garbage cans.

Colourless crystals slowly evolving a heavy vapour of strong odour; slightly soluble in water, readily soluble in organic solvents. Normally used as 100% crystals; stable, non-corrosive and non-staining to fabrics. Toxic effects appear in man at doses above 300 ppm.

Parasite. (*zoo*) A living organism that subsists directly on or within the living tissues of other organisms. For example, the flea (*Xenopsylla cheopis*) on the rat, the hookworm (*Ancylostoma caninum*) in the alimentary tract of the dog, and the larvae of the Cluster fly (*Pollenia rudis*) in the tissues of earthworms. See also ECTOPARASITE.

Parcoblatta. (*ent*) Dictyoptera: Blattidae. A genus of cockroaches (Wood cockroaches), fortuitous invaders of homes in the United States, the females of most species having reduced tegmina and wings, the males flying actively at night. The most widely distributed is *P. pensylvanica* living in woodland, under bark, in hollow trees, wood piles and ground litter. The males are attracted into homes by light; the females are brought indoors with firewood and groceries. Wood cockroaches are unlikely to establish infestations in homes and do not produce the characteristic odour of the pest species.

Paris green. (*chem*) Copper acetoarsenite. An emerald green powder introduced in about 1867 for the control of Colorado beetle and once used for mosquito control; of low solubility in water, but in the presence of water and carbon dioxide decomposes to give water-soluble arsenical compounds. A stomach poison: limited in use to baits and as a mosquito larvicide, now replaced by more effective compounds. Highly toxic to man: acute oral LD_{50} (rat) is 22 mg/kg.

Parthenogenesis. (*zoo*) The ability of females to reproduce without mating (e.g. in some species of booklice).

Particle size. (*phy*) A measure of the fineness of a pesticide formulated as a solid. Fine insecticidal dusts are more effective than coarse dusts: they are more readily picked up by insects, coat the cuticle better and have improved insecticidal action. However, dusts which are too finely ground tend to aggregate because of electrical forces imparted to them during application. They may also increase inhalation hazard. Rodenticidal baits containing the active ingredient as fine particles are also generally more effective: fine particles provide a greater surface area than coarse particles for chemical action or solution in the gut.

The particle size of commercially available dusts is usually less than 150 microns (passing 100 mesh sieve), often as low as 10–40 microns (passing 350 mesh sieve). The particles of AEROGELS (*q.v.*) are sub-micron in size.

Partition wall. (*bldg*) A temporary or permanent structure to separate two parts of a building. The wall common to adjacent properties in, for example, terraced housing. Such walls may be of solid construction (brick as in the above) or may be of fibre or particle board applied to both sides of timber framing. The latter construction provides cavities suitable for rat and mouse harbourage and are best treated with rodenticidal contact dusts. Partition walls between properties are often so thin, or meagre in construction, as to allow easy access of pests from one property to another, e.g. bedbugs.

Parts per million. (*tox*) ppm. The number of parts of a substance (usually by weight) contained in one million parts of a diluent or other substrate. E.g. a measure of the level of pesticide contamination in food, and the 'safe limit' for admixture of insecticides with foodstuffs (e.g. MALATHION 8–10 ppm).

Parturition. (*zoo*) The act of giving birth.

Passer domesticus. (*zoo*) House sparrow. (English sparrow in U.S.). A pest bird in certain food manufacturing industries, notably bakeries and works canteens, where food residues encourage sparrows to feed in large numbers. Problems arise in loading bays and other areas within buildings where products in manufacture and awaiting despatch are fouled by droppings. Employees are often at fault by deliberately feeding the birds. Available food encourages sparrows to nest in the fabric of buildings, the nests becoming reservoirs for insect pests (various beetles, moths and mites). Nests are constructed of twigs and dried grass, on roof supports, ledges and in gutters. Insulation materials are often attacked.

Sparrows move in flocks, they reproduce more rapidly than other pest birds: 3–5 eggs per clutch, 3 broods per year, hatching time 2 weeks, the fledglings spending a further 2 weeks in the nest. Normal food consists of seeds and insects but the entire diet may consist of food residues if housekeeping is neglected.

Control methods in the U.K. consist of proofing and narcotising under licence to the M.A.F.F. See BIRD LAWS.

Paste. (*chem*) A formulation of insecticide or rodenticide usually in a mixture of cereals and oil used as a bait. Examples are pastes of chlordecone or phosphorus squeezed into harbourages for cockroach control; rodenticidal pastes (e.g. of phosphorus or zinc phosphide) are often applied to a carrier, such as bread.

Fig. 58 (*right*) Misting a linen cupboard where wetting of the textiles by heavier spraying may cause damage.

Fig. 59 (*below*) Demonstration of the use of a motorised knapsack sprayer with powder-blowing attachment.

Musca domestica
Fig. 60 (*above*) adult;
Fig. 61 (*centre*) larvae (typical
fly maggots);
Fig. 62 (*bottom*) puparia.

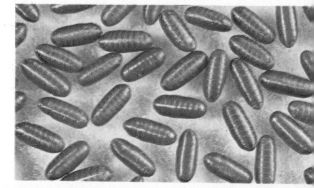

Fig. 63 (*above*) *Mus musculus* in a typical feeding posture.

Fig. 64 (*right*) Nymphs of *Blattella germanica* of mixed age. The adult colouring is not yet developed; the yellow patch on the thorax distinguishes these nymphs from the entirely brown *Blatta orientalis*.

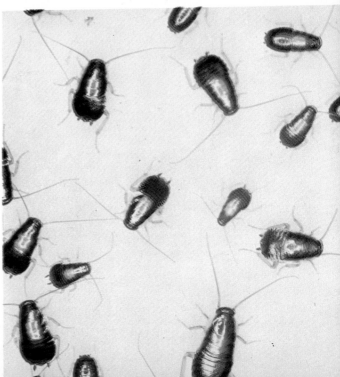

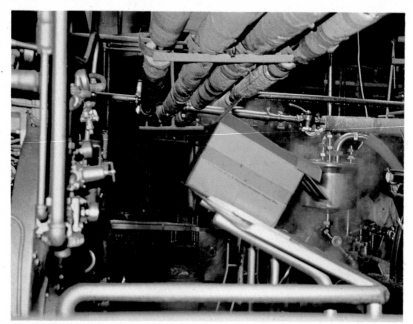

Fig. 65 (*above*) An example of poor practice in food manufacture when packaging materials are allowed into production areas: in this instance a milk processing plant. Note the potential harbourages for cockroaches in steam pipe insulation and hollow metal gantry.

Fig. 66 (*below*) The creation of potential rodent harbourage in a factory perimeter area by the disposal of unwanted machinery and equipment.

Fig. 67 (*top right*) A 'Rentoflash' unit, consisting of an ultra violet light surrounded by charged and earthed grilles, to attract and kill flying insects.

Fig. 68 (*top left*) Use of full protective clothing in the treatment of roof timbers with insecticidal fluid.

Fig. 69 (*below*) A professional pest control serviceman and his equipment.

Fig. 70 (*top*) Adult *Periplaneta australasiae* (female) with oötheca. Note pale 'shoulders' to the tegmina.

Fig. 71 (*centre*) Blocking of gutter and down pipe by bird nesting material (*Sturnus vulgaris*).

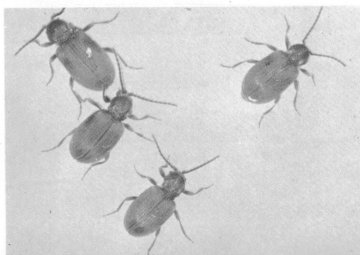

Fig. 72 (*bottom*) *Ptinus tectus:* note the spider-like appearance.

Fig. 73 (*above*) *Rattus norvegicus.* (Kinns).

Fig. 74 (*right*) Rat guards on hawsers to prevent entry by rats to ships in port.

Fig. 75 (*below*) Carriage of goods in trade introduces infestation and may cross-contaminate foods in dockside warehouses.

Fig. 76 (*above*) The accumulation of refuse, encouraging pests outside a food factory.

Fig. 77 (*below*) The correct disposal of refuse, off site, by a local authority, using the principle of "filling and covering".

Fig. 78   Good stacking on pallets with adequate inspection corridors.
(Crown copyright)

Pasturella pestis. (*dis*) The causative bacterium of Plague, primarily a
disease of rodents, but one of the most fatal diseases of man in those
instances that occur (1–2 thousand cases a year). Bubonic plague, the
commonest form of the disease causes large swellings of the lymph
glands. Pneumonic plague causes epidemics in humans, often preceded
by outbreaks among the animal vectors themselves. Urban plague is
usually transmitted from commensal rats by the bite of the Oriental rat
flea (*Xenopsylla cheopis*). Sylvatic plague (the form most commonly
occurring today) is transmitted from wild rodents (gerbils, wood rats,
cotton tail rabbits, pocket gophers and others), also by the bite of various
species of fleas. House mice are seldom involved in the transmission of
plague to man; their fleas are not efficient vectors of the bacterium.
    Fleas carrying the bacterium may infect man in two ways; 1)
regurgitation by the flea when biting its next victim; the bacterium can
multiply in the crop and midgut of the flea and thereafter enter skin
punctures of all the hosts it bites. 2) the bacterium may be dropped
with the faeces of the flea and enter skin abrasions or flea bites.
Pathogen. (*dis*) A disease-producing micro-organism.
PCP. (*chem*) See PENTACHLOROPHENOL.
Pentachlorophenol. (*chem*) PCP, often abbreviated to penta. Introduced
about 1936 for timber preservation, a dark coloured solid with
characteristic odour; non-flammable, non-corrosive but oil solutions
degrade rubber. Used principally as a fungicide (normally at 5%), but
also incorporated with insecticides to give timber preservatives a wider

spectrum of activity. Antiblooming agents are required to prevent the crystallisation of PCP on timber surfaces. Of limited use in termite control.

The acute oral LD_{50} (rat) is 210 mg/kg. Irritating to the skin, mucous membranes and especially to the eyes. Usually used in oil solutions, also as an emulsion cream. See also SODIUM PENTACHLOROPHENATE.

Pentarthrum huttoni. (*ent*) See EUOPHYRUM CONFINE.

Perimeter area. (*bldg*) The area immediately surrounding a building. As a contribution to PREVENTIVE PEST CONTROL the perimeter areas of food factories and other buildings prone to rodent infestation should be under good hygiene control: the surface should be concreted or tar-macadamed so that burrowing is prevented and spilled foodstuffs easily cleared (Fig. 50); alternatively the perimeter area should be landscaped and maintained so that burrows and other evidence of rodent activity can be easily seen; drainage systems must be maintained in good condition; broken gully or manhole covers replaced; outbuildings (painters' and builders' sheds, engineers' stores and garages), from which rodent infestations may spread, should be inspected for good standards of storage, housekeeping and proofing.

The collection and disposal of refuse should be well-controlled (see REFUSE DISPOSAL) and incineration areas well maintained; pallets and boxes stored in perimeter areas should be in orderly stacks and on made-up ground away from walls; old machinery and equipment for disposal should be removed from the premises completely and as quickly as possible (Fig. 66).

Periplaneta. (*ent*) Dictyoptera: Blattidae. A genus containing five pest species of cockroach, all depositing their oöthecae 24–48 hours after formation.

Periplaneta americana. (*ent*) The American cockroach. Thought to have originated in tropical Africa, and now widely distributed throughout the world. It is one of the most important pest cockroaches of the tropics and subtropics occurring in many types of buildings and in warm climates, living outdoors. It is less prevalent in cooler latitudes but infestations develop where artificial environments (e.g. zoos) provide favourable conditions for survival. The temperature preference is 28–32°C.

Indoors the American cockroach is a common pest of restaurants, bakeries, grocery stores and all premises where food is prepared and stored. It is found in basements and higher floors if food is available. In certain areas of the United States it occurs in latrines, privies and sewers. It was once common in the galleys and mess rooms of ships but in these locations the American cockroach has been largely usurped by the German cockroach. *P. americana* is still, however, a frequent traveller among cargoes in holds, notably bananas.

The adult (28–44 mm long) is shining red-brown with a paler yellow area around the edge of the pronotum. The wings overlap the end of the abdomen, more so in the male than the female. The oötheca (8 mm long) is brown-black with 16 teeth along the ridge, usually containing 16 eggs. Oöthecae are produced at 4–7 day intervals and held by the female for about 24 hours before being deposited. Hatching time is about 1 month at 30°C; the number of nymphal stages is variable (usually about 10), the period of development to adult also varying widely with temperature (minimum 5 months). The adult life span may extend to 1 year.

Periplaneta australasiae. (*ent*) The Australian cockroach (Fig. 70), which like *P. americana* is also believed to have originated in tropical Africa, and 'migrated' first in association with slave labour. Among species of *Periplaneta*, the Australian cockroach is second in importance only to *americana* as a world-wide pest of buildings, but is less tolerant of cool temperatures. In infested buildings, *australasiae* occupies a similar habitat to *americana* and in cool climates has occasionally become established in the artificial environment of greenhouses.

The Australian cockroach is slightly smaller (30–35 mm long) than *P. americana*, the wings overlapping the apex of the abdomen in both sexes. The pale edge around the dark pronotum is more distinct, but *P. australasiae* differs from *P. americana* and *P. brunnea* in having pale basal margins to the tegmina (forewings). The oötheca (12–16 mm long) is red-brown and contains about 24 eggs. Development time to adult is not markedly different from *P. americana*.

Periplaneta brunnea. (*ent*) The Brown cockroach, introduced into the U.S. in 1907, is now common in the southern States. It occurs typically in pantries, grocery stores, and outbuildings and has been found in army camps, city dumps, privies and sewers. This cockroach also arrives in Britain on shipments of bananas and two separate infestations have become established in recent years in buildings at London (Heathrow) Airport.

The habitat and appearance of the Brown cockroach is similar to *P. americana*, but the adult is smaller (31–37 mm long), the paired blotches on the pronotum usually less conspicuous, and the tegmina and wings are not usually as long relative to the abdomen. The two species are best distinguished from the shape of the last segment of the cercus. The oötheca is very similar to that of *P. australasiae* containing an average of 24 eggs.

Periplaneta fuliginosa. (*ent*) The Smoky-brown cockroach, a pest of buildings in the southern United States; well established in Georgia, northern Florida and westward to Texas. In the south-east it is common outdoors in garages, outbuildings, woodpiles and porches of houses. It is similar in size (31–35 mm long) and appearance to *P. brunnea*, but is entirely shining brownish-black, almost as dark as *Blatta orientalis*. The wings of both sexes cover the abdomen.

Periplaneta japonica. (*ent*) Yamato-Gokiburi, the Japanese cockroach. A pest cockroach distributed from central to north Japan. The adult

(24–28 mm long) is shining brown-black without pale markings, the wings of the male are well developed, those of the female cover only the anterior half of the body. The oötheca 8–9 mm long, is produced at 6-day intervals and contains 15 eggs. Incubation period of the eggs is 6 weeks (23–29°C). Development of nymphs to adult; 6 months.

Persistence. (*chem*) Of pesticides, remaining active for a long period of time after application; having a long life on treated surfaces (see RESIDUAL INSECTICIDE). Organochlorine compounds (e.g. DDT, dieldrin) are described as persistent insecticides: they are not readily degraded, but may remain or accumulate in some component of the environment over a period of years.

Perthane. (*p. prod*) An organochlorine compound with no common name, related to DDT, introduced as an insecticide in 1950 by Rohm & Haas Co. Recommended for domestic use against clothes moths and carpet beetles, often in combination with other insecticides. Virtually non-toxic to man.

The technical product is a wax, melting at about 50°C; practically insoluble in water, but soluble in most aromatic solvents. Its stability is similar to DDT but less insecticidal.

The acute oral LD_{50} (rat) is 8,200 mg/kg. Formulations include wettable powder, emulsion concentrate and oil solution.

Pest. (*zoo*) A troublesome or destructive animal or thing, from the Latin *pestis* meaning plague. At one time implying the presence of an undesirable organism in large numbers. Now any living organism (plant or animal) which is detrimental to man or his activities in some way: by example, one fly in a high class restaurant can cause damage to the reputation of the establishment, the number of flies not necessarily being important.

Pest control. (*proc*) The application of techniques, principally the use of chemicals, in protecting the health and well-being of man and domestic animals; the elimination of disease-carrying organisms; the protection of foodstuffs and property from damage and economic loss; the prevention of contamination of foods by biological organisms and their metabolic products; the improvement of man's comfort and the elimination of animals which are repulsive, cause fear or annoyance. The requirements are primarily, 1) the ability to recognise pests, 2) basic knowledge of their habits, and 3) recognition of the hazards and safe use of pesticides.

In addition to killing and trapping, measures of preventive pest control make a significant contribution. These are methods of denying pests entry into building structures (see PROOFING) and of environmental control aimed at removing the conditions which pests need to breed and grow (see SANITATION).

Pest control contract. (*name*) An undertaking by a pest control contractor to eradicate a pest, in or around a client's premises, without putting his client or customers at risk and on agreed terms, specifying principally the duration of the contract, the number of visits to be made and the price for the work. It is the contractor's job to provide no less than has been agreed, to undertake more treatments if necessary, to use the most efficient pesticides available, conforming always with national or local legislation on pesticide use, and respecting always the safety of personnel, the client and his property.

It is not uncommon for contracts to include action by the client, such as PROOFING (*q.v.*) and improved HOUSEKEEPING (*q.v.*) in assisting the activities of the pest control contractor.

Some pest control companies offer guarantees, the period varying from 6 months (perhaps following one treatment for cockroach control), to 20 years (for the eradication and protection of timbers against beetle attack).

Pesticide. (*chem*) A substance which kills pests, usually toxic also to non-pest organisms. See SPECIFICITY.

Pesticide analysis. (*proc*) See ANALYTICAL METHODS.

Pesticides Safety Precautions Scheme. (*name*) Originally the Notification of Pesticides Scheme, changed to the Pesticides Safety Precautions Scheme (PSPS) in 1955. The means for regulating the use of pesticides in the U.K. Unique in its flexible operation through agreement with the major chemical and servicing companies, combined with limited legislation for certain chemicals and their uses.

To observe the scheme, a manufacturer, importer, distributor, or servicing operator, who wishes to introduce a new chemical, or new use of an existing chemical, must notify it to the M.A.F.F. The onus is upon the notifier to provide all the information required to satisfy government departments that the chemical will be safe when used in the manner proposed.

Information submitted is considered by the Scientific Subcommittee of the Advisory Committee on Pesticides and other Toxic Chemicals. The Scientific Subcommittee is composed solely of scientists with expert knowledge of some aspect of pesticides. The Advisory Committee consists of representatives, administrative, scientific and technical, of interested departments with other independent members. Neither committee contains trade representatives. Small groups of experts (panel members) make contributions in particular fields, e.g. medical, operator protection, residues and wildlife.

Draft recommendations agreed by the committees are cleared with the notifier and published as 'Chemical Compounds used in Agriculture and Food Storage—Recommendations for Safe Use in the U.K.'. The advice given is in three sections dealing with risks to the operator, consumer and wildlife.

The great advantage of the Scheme lies in its flexibility: the record of the high safety standard achieved as well as the respect and communication developed between Government, industry and conservation bodies,

stands as a tribute to voluntary procedures. See Legislation, Approval
of Pesticides and Labelling.

Pesticides Survey Group. (*name*) Two Pesticide Survey Groups were established in the U.K. in 1965: one at the Plant Pathology Laboratory,
Harpenden, dealing with agricultural and horticultural uses of pesticides, the other at the Pest Infestation Control Laboratory, Slough,
concerned with pesticide use in food storage practice, home kitchen and
larder, and certain industrial applications. Both were set up following a
recommendation of the Advisory Committee on Pesticides and Other
Toxic Chemicals. Their main functions are to obtain qualitative and
quantitative data on the uses of pesticides in the U.K. and to publish
the results of their surveys. A summary of work carried out by both
groups to the end of 1968 was published as an appendix to the 'Further
Review of Certain Persistent Organochlorine Pesticides Used in
Great Britain (1969)'. The Department of Agriculture and Fisheries
for Scotland has also made arrangements for similar data to be
obtained.

Pest Infestation Control Laboratory. (*name*) P.I.C.L. Formed in 1970 by a
merger of the Ministry's Infestation Control Laboratory (M.A.F.F.)
and the Agricultural Research Council's Pest Infestation Laboratory.
The Laboratory provides technical and scientific support to the Ministry
of Agriculture, Fisheries and Food and if necessary to other government
departments through research on the storage of grain and other commodities: in particular, pest control by physical means, insecticides and
fumigants; rodent control and rodenticides and the control of other
mammals and birds in agriculture. The work also covers relevant aspects
of public health. Non-agricultural uses of pesticides, pesticide residues
and the effects of pesticides on wildlife are also studied.

The Laboratory occupies three main sites: Slough (food storage and
storage pests); Tolworth (rodents and vertebrate biochemistry) and
Worplesdon (other mammals and birds). The principal work done for
other departments is research on the bird-strike hazard to aircraft.
Addresses of the main Laboratories:

London Road, Slough, Bucks. SL3 7HH.

Govt. Buildings, Hook Rise South,
Tolworth, Surbiton, Surrey KT6 7DX.

Tangley Place, Worplesdon,
Guildford, Surrey.

Pharaoh's ant. (*ent*) See Monomorium pharaonis.

Pharate. (*ent*) The condition of a larva, pupa or adult remaining enveloped
by the cuticle of the previous stage. E.g. the pharate pupa of *Musca*,
so described because the last larval cuticle is never shed, remaining as
the puparium.

Pharmacology. (*tox*) The science of drugs and their action.

Phauloppia lucorum. (*zoo*) Acari: Oribatidae. A mite often invading properties in large numbers in summer, principally in the roof void, spreading throughout the house causing annoyance but no damage. Brown-black, with a hard shell-like structure covering the body. Food consists of vegetable matter; lichens and algae.

Pheromone. (*chem*) A substance produced by insects in sub-micro amounts providing a means of 'communication' between individuals (e.g. trail pheromones of ants, AGGREGATION PHEROMONES (*q.v.*) of cockroaches). Also between the sexes as a stimulus to mating. See also ATTRACTANT.

Phosphine. (*chem*) See ALUMINIUM PHOSPHIDE.

Phosphorus. (*chem*) Yellow phosphorus. A white-yellow waxy solid almost insoluble in water, igniting spontaneously in air at 30°C and above. Once used extensively as an acute poison; as a paste smeared on bread for rodent control, and as an oily bait for cockroach and cricket control. Now largely replaced by safer compounds; banned for rodent control in the U.K. by the ANIMALS (CRUEL POISONS) ACT (*q.v.*). Fatal dose to man is about 50 mg.

Phostoxin. (*p. prod*) See ALUMINIUM PHOSPHIDE.

Physical control methods. (*proc*) Techniques of pest control not involving the use of chemicals with toxic properties, e.g. the Entoleter (Fig. 79), mechanical traps and sticky boards. Methods include the use of ultrasonics (for repelling rodents) and ultra violet light (for fly attraction, Fig. 67).

Physiology. (*zoo*) The science of the processes of life. The study of body functions; nutrition, excretion, respiration, sensory perception, coordination, behaviour, reproduction and the regulation of body functions in response to the environment.

Phytotoxic. (*tox*) Toxic to plants. In outdoor applications of insecticides (e.g. against CASUAL INTRUDERS (*q.v.*)) wettable powders are less likely to cause leaf scorch than emulsions and oil sprays.

Pied Piper. (*name*) A legendary figure of German folklore who by the attractive properties of his flute lured rats from the town of Hamelin in Saxony, 600 years ago, to drown in the River Weser. Not being paid for this service he led all the children away into a cave which closed after them and they were never seen again.

Various explanations of this legend have been postulated, the latest being that it is an account of a visitation by the Plague; two outbreaks apparently occurred in Hanover in 1348 and 1361. Carried by rat fleas, the disease would kill a great number of the rats and perhaps naturally the citizens would have collected the corpses and disposed of them in the river. Assuming that the incident refers to the second outbreak, a good deal of immunity may have been conferred on the human survivors of the first, but not to children under thirteen. With death accounting for so many children it would have been natural to arrange a mass burial

Fig. 79 Entoleters in a flour mill. A physical method of pest control by which insect eggs are destroyed by centrifugal force. (H. Simon Ltd.)

in a local hillside. This together with the common practice in Medieval art of depicting Death as a dancing skeleton leading people to their final fate may account for the 'queer long coat from head to heel, half of yellow and half of red'.

The town is now kept free from rats by the German subsidiary of Rentokil. With the legend in mind this Company has a contract which provides for payments quarterly in advance.

Pigeon. (*zoo*) See COLUMBA LIVIA.

Pillbug. (*zoo*) See ISOPODA.

Pindone. (*chem*) An indane-dione type anticoagulant rodenticide, first shown to have insecticidal properties in 1942, originally suggested as an alternative to pyrethrum in fly sprays, but subsequently (1944) introduced as a rodenticide by Kilgore Chemical Co. under many trade names, including 'Pival'.

The technical material is a crystalline solid, bright creamy-white with slight odour; virtually insoluble in water, but soluble in dilute alkalis, ammonia and most organic solvents.

The acute oral LD_{50} (rat) is 280 mg/kg. More toxic when administered in small daily doses. Somewhat less toxic and less palatable to rats than warfarin at 0·025%. Dogs are more susceptible than rats (acute oral LD_{50} 74–100 mg/kg) and are killed by daily doses of 2·5 mg/kg.

Formulations available include a 0·5% powder, for admixture in rodent baits at 0·025%, and the sodium salt ('Pivalyn') used as a liquid bait.

Pipe chasing. (*bldg*) See DUCT.

Piperonyl butoxide. (*chem*) The most widely used synergist of pyrethrins and synthetic pyrethroids, increasing knockdown, toxicity and persistence; discovered in 1947 (see SYNERGISM). Used in combination with pyrethrins and related compounds at ratios varying from 5 to 20 parts piperonyl butoxide to 1 part pyrethrins by weight, in aerosols, oil-based formulations, emulsions and dusts.

A pale yellow odourless oil, virtually insoluble in water, soluble in most organic solvents, stable to light and non-corrosive. Acute oral LD_{50} (rat) is 7500 mg/kg. May be irritating to mucous membranes.

Pival. (*p. prod*) See PINDONE.

Pivalyn. (*p. prod*) See PINDONE.

Plague. (*dis*) See PASTURELLA PESTIS and XENOPSYLLA CHEOPIS.

Plant extracts. (*chem*) See BOTANICAL INSECTICIDES.

Plaster beetles. (*ent*) See FUNGUS BEETLES.

Plastic strip curtain. (*equip*) Lengths of plastic material hung in doorways and other openings, the strips reducing the effective space for entry of flying insects, but not impeding the movement of personnel. Available also as overlapping strips (40 cm wide) mounted on slide rails and sufficiently robust to allow the impact of moving vehicles, forklift trucks and goods on conveyors, but in the 'rest' position closing the opening against insects and draughts. See also FLY SCREENS and AIR CURTAIN.

Plinth. (*bldg*) The recessed front panel at floor level of storage units (such as cupboards and cabinets) as found in kitchens, laboratories and quality control rooms. These units invariably have false bottoms and provide rodent and insect harbourage. Such cabinets should have the front panel —the plinth—removed, or fitted in such a way that it is possible to inspect the void. Ideally storage units of this type should be on legs, with the base not less than 20 cm off the floor.

Plodia interpunctella. (*ent*) Lepidoptera: Pyralidae. The Indian meal moth. A pest of imported cargoes, especially nuts, dried fruit and grains. The common name is said to come from the U.S. because of the association of this insect with maize meal (Indian corn). The larvae are very pale, usually yellowish white. The adult flies actively, the shoulder areas of the forewings are pale buff, separated from an outer red-brown area by a dark brown line. The minimum period from egg to adult is 6 weeks but is much extended at lower temperatures and on poor diets.

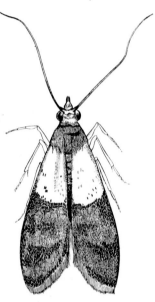

PMP. (*chem*) See VALONE.

Poison. (*chem*) A toxic substance. Commonly, implying any chemical which, when taken into the body of man, animals or into plants, impairs health or kills.

Poisons cupboard. (*equip*) A strong locked cupboard, or box marked 'Poisons', containing the more toxic pesticides (e.g. fluoracetamide, sodium monofluoroacetate, thallium sulphate, Cymag and other compounds as may be felt necessary); an essential provision within a pesticide store. Also a 'Poisons book' which records the date, name of pesticide, quantity entered, withdrawn and stock remaining, the name of the person withdrawing and the location on which used. The object is to provide special security, limited access to pesticides of high toxicity and records of use.

Poisons Rules. (*leg*) Regulations conforming with the Pharmacy and Poisons Act (1933) and revised in 1970, restricting by law the availability, sale and use of scheduled poisons in the U.K. Relevant to pest control in controlling the purchase and use 1) of strychnine for killing moles and 2) of sodium monofluoroacetate and fluoracetamide as rodenticides for use in ships, and sewers, warehouses and in drains (situated in restricted dock areas of ports).

Pollenia rudis. (*ent*) Diptera: Muscidae. Cluster fly. Migrates indoors and into roof spaces, in the autumn, often in vast numbers, emitting a sickly sweet smell. A nuisance pest, sluggish in flight, congregating in summer on outdoor walls and other structures.

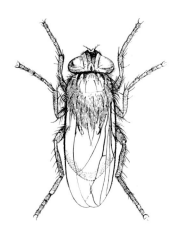

The adult (8 mm long) is dark, grey-olive, the thorax with densely matted golden hairs. The wings almost completely overlap at rest. Eggs are laid under dead leaves; the larvae parasitise earthworms and pupation occurs in the soil. Probably only one generation a year in Britain, but several summer generations occur in warmer climates. See CLUSTER FLIES.

Pollution. (*chem*) See CONTAMINATION.

Population. (*zoo*) Of pests, the number of individuals comprising an infestation; many thousands, or millions, of insects in stored foodstuffs; hundreds of pigeons roosting on a building; perhaps only 10 or 20 rats harbouring in a factory.

Post-construction treatment. (*proc*) See IN-SITU TREATMENT.

Potentiation. (*tox*) An increase in the mammalian toxicity of a chemical by admixture with another as when two insecticides are mixed (e.g. Gardona with dimethoate, malathion or carbaryl) cf. SYNERGISM (*q.v.*)—an increase in insecticidal activity.

Poultry red mite. (*zoo*) See DERMANYSSUS GALLINAE.

Powder. (*chem, phy*) See AEROGEL, DESICCANT, DIATOMACEOUS EARTH, DUST, PARTICLE SIZE and WETTABLE POWDER.

Powder post beetle. (*ent*) See LYCTUS BRUNNEUS.

ppm. (*tox*) See PARTS PER MILLION.

Pralidoxime chloride. (*chem*) A specific antidote for use in conjunction with ATROPINE SULPHATE (*q.v.*) for the treatment of poisoning by organophosphorus insecticides; registered under the trade name 'Protopam chloride', also known as '2 PAM' and 'PAM'. Administered intravenously, or in mild poisoning cases by tablet. Its action is to reactivate cholinesterase. Dose: 1 g for adults; 0·25 g for children.

Pre-baiting. (*proc*) The practice of laying unpoisoned food to encourage feeding at particular locations so that toxic baits, laid subsequently, are more readily eaten. A necessary preliminary to the use of many acute rodenticides, also for improving the efficiency of bird trapping and narcotising. It is best that the food used for pre-baits be the same as that proposed for the toxicant, or for trapping purposes.

Precautionary measures. (*leg*) See LABELLING.

Predacious. (*zoo*) Living by preying on another: the commonest predators in buildings are the carnivorous beetles (members of the Carabidae) and mites (e.g. *Cheyletus eruditus*).

Premium grade. (*chem*) See TECHNICAL MATERIAL.

Preventive pest control. (*proc*) Measures taken by management, especially those concerned with catering, food manufacture and warehousing, to deny the entry of pests and their establishment in buildings. The objectives are, 1) to make buildings inaccessible or unattractive to pests and 2) to provide conditions within, which allow the easy and swift detection of pests should they gain entry, allowing eradication to be carried out easily.

Many different activities make a contribution to these objectives; among the most important are attention to PERIMETER AREAS, PROOFING, inspection of incoming materials from SUPPLIERS, TRANSPORT and depots, and RETURNED GOODS. They include also considerations of building structure, cleaning, HOUSEKEEPING, STORAGE and above all management attitude.

Primary reproductive. (*ent*) A caste of termites: winged males and females which establish new colonies. Usually darker than other members of a termite colony. Swarming usually occurs in spring and autumn when the reproductives leave the nest in large numbers; often the first sign that termites are present.

Processed food. (*manuf*) A raw edible material which has undergone some change in manufacture, usually involving heat, to convert it into a more desirable product (e.g. breakfast cereal). If contaminated by pesticides, processed foods present a greater health risk, because there is no further manufacturing stage to reduce the pesticide level by dilution or degradation.

Production area. (*manuf*) Part of a factory in which the manufacturing processes are housed.

Pronotum. (*ent*) The upper or dorsal surface of the prothorax of an insect, often enlarged, forming a shield-like structure partly covering the meso- and metanota, (e.g. in cockroaches) or extending over the head (e.g. in *Rhizopertha dominica*).

Proofing. (*proc*) Measures taken to reduce or eliminate entry points for pests into buildings (Fig. 80) and the movement of pests between departments of buildings. Examples of access points for rodents include: ill-hung doors, uneven floors beneath doors, particularly roller shutters, damaged doors and frames, missing mortar fillets between door jambs and brickwork; gaps between sliding doors and frame uprights, and gaps around pipework passing through external walls. The same applies within buildings but additionally, holes left in walls and floors where electrical conduit and other pipework has been removed or newly installed.

All such gaps and holes should be effectively sealed. Apart from ensuring that doors fit well, metal plates should be fitted to the toes of doors and as cladding to the bottom of frames; compressed wire netting should be forced into openings before cementing cracks and gaps around pipes in brickwork thereby making a secure joint; caulking mastic should be used to fill expansion joints on external surfaces; air bricks, vents and other small openings should be proofed with materials not likely to corrode.

In the interests of weather proofing, good warehousing practice demands immediate repair of broken windows, damaged roofs and doors. Stores should be insulated against heat and damp and warm service pipes insulated to prevent local overheating. See also FLY SCREENS and BIRD PROOFING.

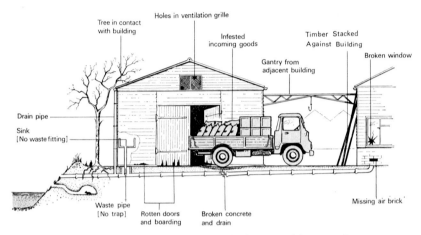

Fig. 80 Points of entry of rodents into buildings requiring proofing measures and inspection of incoming goods.

Propellent. (*chem*) The component of an aerosol product, e.g. domestic fly killer, providing the pressure to make the contents self-dispensing when the valve is depressed. May be 1) a gas under pressure (e.g. nitrogen, nitrous oxide or carbon dioxide), or more often 2) a liquid under pressure (e.g. butane, propane, or fluorochlorinated hydrocarbons) which becomes a gas as the pressure is released (e.g. the proprietary products, Arcton, Freon and Isecon). Compressed liquids vaporise as the AEROSOL (*q.v.*) is emitted and often act as solvents or co-solvents for the active ingredient(s).

Propoxur. (*chem*) Arprocarb. A carbamate insecticide introduced in 1959 by Bayer under the trade names, 'Baygon' and 'Blattanex', of value for the control of cockroaches, ants, flies, mosquitoes and other household insects. CROSS-RESISTANCE (*q.v.*) with diazinon in cockroaches is not evident.

The technical product is a crystalline powder with a faint characteristic odour, sparingly soluble in water and not readily soluble in conventional organic solvents. Knockdown is rapid; some 'flushing' action for cockroaches is apparent.

The acute oral LD_{50} (rat) is 100 mg/kg. It is highly toxic to bees. Formulations available include emulsion concentrate, wettable powder, dust, bait and aerosol.

Protected species. (*zoo*) Animals fostered by legislation, shielded from extinction or reduction in numbers, because of their aesthetic contribution to man's environment. Animals against which pest control operations are forbidden by law (e.g. robins and chaffinches in the U.K.; see BIRD LAWS).

Protective clothing. (*equip*) Dress and equipment for self-protection against the toxic effects of pest control chemicals (Fig. 68), consisting for general pest control work of, boiler suit, dust coat (regularly laundered) and GLOVES (*q.v.*); Wellington boots (for outdoor work) and cotton over-shoes (for certain applications in domestic properties). Also a GAS MASK (*q.v.*) to protect against fluid droplets or vapour, and a DUST MASK (*q.v.*) to prevent inhalation of toxic particles. Those involved in fumigation should be specially equipped with masks and appropriate canisters to prevent the inhalation of toxic gases.

The informed pest control operator is aware that the day-to-day application of pesticides may put him at risk to the chronic effects of pest control chemicals and that an occasional mishap may subject him to the risk of acute poisoning. However remote these risks may be, he knows that it is wise to adopt precautions by wearing protective clothing. Also that it enhances his appearance as a professional man, and raises in the eyes of the public and his clients the responsible attitude adopted by the pest control industry towards health and safety.

Prothorax. (*ent*) The foremost segment of the thorax of an insect immediately behind the head and bearing the first pair of legs. In most insects (e.g. cockroaches and beetles) the dorsal surface of the prothorax (the pronotum) is much enlarged hiding the other thoracic segments from view.

Protopam chloride. (*p. prod*) See PRALIDOXIME CHLORIDE.

Psocoptera. (*ent*) Booklice, sometimes called Psocids. Common resident pests in libraries, homes and warehouses; often occurring in stored foods. Psocids feed principally on microscopic moulds on damp materials (glues, size, books, paper, framed pictures, wallpaper, woodwork, leather and plaster). They also live on foods with a high vitamin B content. A common species is *Liposcelis divinatorius*. Small insects (1–2 mm) which run rapidly, with soft bodies, cream to light brown. Wings sometimes absent, the thorax very small compared with the rest of the body; females of some species reproducing without mating (parthenogenesis).

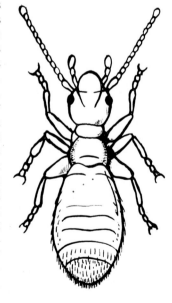

The eggs, laid separately, are sticky and adhere to food; there are 3–8 nymphal stages according to the species;

the life cycle may be as short as 25 days. Adults may live for 6 months.

Booklice are common in newly built properties whilst the plaster is drying out. Ventilation and drying prevents the growth of moulds and thus eradicates an infestation. Otherwise, control moulds with SODIUM PENTACHLOROPHENATE, separately or combined with an insecticide. Valuable books and pictures should be fumigated.

Psychodidae. (*ent*) Filter flies or Owl midges. Tiny flies, about 2 mm long, abundant in spring and summer, often becoming a nuisance to properties near sewage works. The adult flies are grey with hairy wings held roof-like over the body when at rest. The eggs are laid in batches of up to 200 on wet decaying matter, particularly the filters of sewage beds. The larvae, legless and slow moving, live almost submerged in the sludge on which they feed, helping to increase the rate of breakdown of the sewage material. Filter flies may have as many as eight generations per year. Treatment of Filter flies is complicated by the presence in the filters of bacteria essential for the decomposition of organic matter. The treatment of fly alighting surfaces and fogging, to knockdown large numbers of flying adults, can however bring local relief where the flies are troublesome.

Ptinidae. (*ent*) The family of the Coleoptera containing the spider beetles; adults with a waist-like constriction at the back of the thorax and rather long legs and antennae, imparting a resemblance to spiders. See PTINUS, GIBBIUM, NIPTUS and TRIGONOGENIUS.

Ptinus tectus. (*ent*) Coleoptera: Ptinidae. The Australian spider beetle. A widespread pest of stored products, occurring in temperate regions only. An established pest of flour mills. A scavenger on all vegetable and animal debris including rat and mouse droppings and old birds' nests. Causes damage to cereals and spices; also bores into wood and books.

The adult (3–4 mm long) is dull red-brown with a constriction between thorax and abdomen and with long legs and antennae, giving a spider-like appearance (Fig. 72). The elytra are densely covered with brown hairs.

The minimum period of development from egg to adult is 2 months, more often 3–4 months. Adults live up to one year and feign death when disturbed.

Public Health Inspector. (*name*) An officer appointed to protect the health of the public. Established in England and Wales by the Public Health Act of 1848, which set up local Boards of Health, subordinate to a General Board. The Act provided a system whereby local matters of health came under the control of central government: the stimulus came from alarm at the time about the heavy incidence of disease throughout the country, and a demand that knowledge of disease prevention be applied for the protection of public health, in the form of 'legislative measures of sanitary reform'.

The 1848 Act required the appointment by all local Health Boards, of fit and proper persons to be 'Inspectors of Nuisances'. These local

Boards were further required to make bye-laws for regulating the duties of these officers which in 1855 became known as 'Sanitary Inspectors'. Since changed to 'Public Health Inspectors'.

The duties of a Public Health Inspector cover a very wide field: the following are examples. He shall:

1) perform all duties imposed by statute, and any regulations made by the Minister of Health and by any local bye-laws.
2) keep himself informed of sanitary circumstances in his district in respect of 'nuisances that require abatement'.
3) report to the local authority any 'noxious or offensive businesses, trades or manufactories'.
4) report any damage to water works or water wastage; 'the fouling by gas, filth or otherwise' of water for domestic purposes.
5) visit and inspect shops and places kept, or used, for the preparation or sale of any article of food.
6) carry out the duties of a sampling officer under the Food and Drugs Act, 1938.
7) give notice to the Medical Officer of Health of the incidence of any infectious or epidemic disease or other serious outbreaks of illness.
8) if directed by the local authority, act as officer for the authority under the Rats and Mice (Destruction) Act, 1949.

Public relations. (*name*) In pest control, the use of all available media to promote: 1) public understanding of pests and their effects on man and his well-being, 2) the services and products available to deal with pest problems, 3) the rational presentation of facts concerning pesticides and their effects on man and the environment and 4) the responsible attitude of the pest control industry toward national and local legislation on pesticide use, and the industry's contribution in helping to frame such regulations.

Pulex irritans. (*ent*) Siphonaptera: Pulicidae. Human flea. The primary host is man, but sometimes occurring on domestic and some wild animals. Widely distributed, once common pests of slum and other poor properties. The prevalence of fleas has declined with the adoption of higher standards of hygiene. Not normally carriers of disease.

Larvae feed in crevices in flooring, on organic matter, including undigested blood eliminated in the faeces of adults. Pupae are formed in silken cocoons incorporating particles of dirt; emergence from pupae may be delayed until suitable hosts for the adults become available. Adults hide in bedding and upholstered furniture (notably theatre and cinema seats). Both sexes bite and one flea will bite many times, activity being concentrated round the waist and upper parts of the body.

Development from egg to adult under warm conditions takes about 1 month.

Pulse. (*plant, zoo*) Collectively, beans, peas, lentils and the seeds of other leguminous plants used as food. Also describing the plants which produce these seeds. In zoology, the beating of the heart and the movement of blood through the arteries.

Pup. (*zoo*) A newly born rat or mouse, hairless, blind and dependent on the mother for food.

Pupa. (*ent*) pl. Pupae. The third stage of development, following the larva, of an insect with complete metamorphosis, often enclosed in a cocoon made by the last larval instar. The pupa is frequently referred to as a resting or quiescent stage; it is incapable of feeding, and is usually limited in movement, although some are adapted for active locomotion (e.g. mosquitoes). Internally, however, there is active reorganisation of the larval body and its organs into those of the adult. In most insects, a pre-pupal larva or PHARATE pupa intervenes between the last larval instar and the pupa. In the Lepidoptera, the pupa is known as a chrysalis.

Puparium. (*ent*) The last larval skin which is retained as a hardened covering for the pupa within (as in certain flies, e.g. *Musca*, Fig. 62).

Pybuthrin. (*p. prod*) Pyrethrins synergised with piperonyl butoxide. Introduced by Cooper, McDougall and Robertson in the early 1950's.

Pyrethrins. (*chem*) A mixture of Pyrethrins I and II and Cinerins I and II; esters occurring in various proportions in pyrethrum extract. One of the most useful insecticides, well-known for 1) its rapid action, such that the insect is either immobilised (see KNOCKDOWN) or 2) repelled from the treated area (see FLUSHING AGENT), 3) wide spectrum of activity against many insect pests, 4) safety in the presence of foodstuffs, and 5) lack of toxic residues in man. Pyrethrins however have a number of disadvantages: 1) harvesting, processing and transport combine to make them expensive, 2) they are not particularly effective as killing agents but when used in combination with a SYNERGIST (*q.v.*) there is a marked increase in efficiency, and 3) very short residual life on treated surfaces.

Uses are varied, but principally 1) the control of flies, mosquitoes and stored food pests, and 2) as a flushing, or knockdown agent, in combination with other insecticides to increase speed of kill.

The technical product is a brown resinous oil, insoluble in water, soluble in many organic solvents; rapidly inactivated in air and decomposed on exposure to light. The acute oral LD_{50} (rat) is 1500 mg/kg. Irritating to the eyes; contact with the skin may cause allergic reactions in susceptible subjects. Formulations are many and varied, including aerosols, sprays and dusts.

Pyrethroids. (*chem*) See SYNTHETIC PYRETHROIDS.

Pyrethrum. (*plant*) *Chrysanthemum cinerariaefolium*. The plant from which PYRETHRINS are extracted (see also PYRETHRUM EXTRACT); mainly the mature flower heads, the achenes being the principal site of concentration.

Pyrethrum extract. (*chem*) A mixture of extracts, only some of which are insecticidal, obtained from the Pyrethrum plant. See PYRETHRINS. Extracts have replaced the earlier use of dried flower heads. The pyrethrin content of extracts ranges from 1·3–3%. Usually available as concentrated extracts of 25%.

Q

Quick. (*p. prod*) See CHLOROPHACINONE.

R

Rabies. (*dis*) Hydrophobia. Caused by a virus in the saliva of an infected animal (man, dog, bat or other mammal); transmitted by a bite. Characterised by nervous derangement, excitability, irritability and lack of consciousness passing into paralysis. Met with in all continents except Australia; largely stamped out in the U.K. by legislative measures involving quarantine. A paralytic form of rabies occurs in the West Indies and S. America, spread by the Vampire bat (*Desmodus rufus*); many species of insectivorous bats are implicated in the United States. Rats have been shown not to be involved.

Rabon. (*p. prod*) See TETRACHLORVINPHOS.

Racumin. (*p. prod*) See COUMATETRALYL.

Radione. (*p. prod*) An anticoagulant rodenticide of the indane-dione type marketed by Lyonnaise Industrielle Pharmaceutique; of lower palatability to *Rattus norvegicus* than warfarin and less effective as a rodenticide.

Rat block. (*chem*) See BAIT BLOCK.

Rat guard. (*equip*) A piece of galvanised metal, which acts as a barrier to the climbing activity of rats, as around brick or concrete pillars at sub-floor level. Also, as round discs on the hausers of ships in port (Fig. 74).

Raticate. (*p. prod*) See NORBORMIDE.

Ratilan. (*p. prod*) See COUMACHLOR.

Rats. (*zoo*) See RATTUS NORVEGICUS and RATTUS RATTUS.

Rattus norvegicus. (*zoo*) Brown rat, Common rat, Norway rat, Sewer rat (Fig. 73). A world-wide pest of industrial, commercial and domestic properties. Omnivorous, but with a preference for cereals. Distinguished from *R. rattus* by the heavier, thick-set body, blunt snout, smaller ears, and the tail which is shorter than the head and body. Colour is variable. *R. norvegicus* usually lives in burrows in soil; railway embankments, canal sides, beneath buildings, infesting drains and sewers, and rubbish dumps. It may live entirely within buildings, where the range of movement is limited, travelling much further outdoors. Agile and a good climber but less so than R. RATTUS (*q.v.*).

Mating is frequent; the gestation period is about 3 weeks, average litter is of 7–8 young; the number of litters a year is 3–6, the period from birth to sexual maturity is 10–12 weeks. Very few Brown rats live more than a year. The maximum number of young produced by a pair and their offspring in a year is about 200. For pest status of rats, see RODENT CONTROL.

Rattus norvegicus has developed resistance to warfarin in the U.K., the major resistance outbreaks are on the Montgomery–Shropshire border and an area N.E. of Glasgow; other small scattered outbreaks have been detected but their extent is ill-defined. Resistance is also present in Jutland and recent reports have come from the N.E. United States.

Rattus rattus. (*zoo*) Black rat, Ship rat. Distributed world-wide but limited in the U.K. mainly to port areas. A major pest of ships. Omnivorous, but with a preference for fruits and vegetables. Distinguished from *R. norvegicus* by the more slender body, pointed snout, larger ears, and tail which is longer than the head and body. Colour is variable. The Black rat is an agile climber generally moving high up in buildings; frequently using beams and other structures as runways, often using trees and connecting wires to enter roof spaces. The range of movement is extensive both within and outside buildings.

The reproductive characteristics are similar to *R. norvegicus*; but the Black rat is more difficult to control because of its food preferences and wide foraging activities.

There are several subspecies of *Rattus rattus*, all very variable in colour which is not reliable for purposes of identification, *viz. R.r. rattus* is the Black rat or Ship rat; *R.r. alexandrinus* is the Grey-bellied or Alexandrine rat; and *R.r. frugivorus*, the White-bellied or Roof rat. The last is a pest of sugar cane fields in Guyana and shows resistance to anticoagulant rodenticides.

Ravap. (*p. prod*) See TETRACHLORVINPHOS.

Raviac. (*p. prod*) See CHLOROPHACINONE.

Red mite of poultry. (*zoo*) See DERMANYSSUS GALLINAE.

Red spider mites. (*zoo*) See BALAUSTIUM and BRYOBIA.

Red squill. (*chem*) The powdered or liquid extract of the dried bulb, *Urginea maritima*, containing scillirocide, used as an ACUTE POISON (*q.v.*) principally against brown rats. Bitter tasting; moderately slow acting. Said to be a powerful emetic in most animals and therefore of low hazard to cats and dogs.

The liquid extract is soaked into cubes of bread, or other bait base, with added flavouring. The powdered extract has been used as a CONTACT DUST (*q.v.*) against house mice.

The acute oral LD_{50} (rat) is 500 mg/kg. The powder is extremely irritating to the skin and causes violent sneezing. Red squill is banned from use in the U.K. by the ANIMALS (CRUEL POISONS) ACT, 1962 (*q.v.*).

Refuse disposal. (*proc*) The removal of all unwanted materials from commercial, industrial and domestic properties, usually the responsibility of local authorities under the supervision of health departments.

Frequent collection and disposal of waste makes a major contribution to PREVENTIVE PEST CONTROL (*q.v.*). On factory sites, refuse containers should be 1) sufficient in number to take the maximum amount of refuse, 2) securely lidded and in sound repair, 3) washed out regularly to remove caked deposits, 4) sited on made up ground which can be easily cleaned, 5) near a water point for hosing down, 6) kept off the ground to make cleaning of the area easy, unless bulk containers and contractors refuse containers are used. (Not as shown in Fig. 76.) Thus, bottles, drums, casks, barrels and similar containers, should be washed out immediately they become empty and before being disposed of. If waste is collected by a number of contractors it is good practice to sort waste into categories; wet, dry, edible, non-edible, combustible and non-combustible, using different bins in the refuse compound.

Those concerned with preventive pest control would be well-advised not to establish refuse tips on their own land. It is far better to have refuse removed to local authority tips (Fig. 77) where any pest problem will not directly threaten manufacturing, storage or catering operations. Refuse should be cleared from factory sites daily. Incinerators are not recommended unless 1) the installation is adequate to cope with the quantity of refuse to be disposed of (i.e. consumes one day's refuse in one day) and 2) the incineration area can be thoroughly cleaned each day.

Registration of pesticides. (*leg*) See APPROVAL OF PESTICIDES and LABELLING.

Relative humidity. (*phy*) The amount of water in the air expressed as a percentage of the maximum (at air saturation), at the prevailing temperature and pressure. Measured with a wet and dry bulb thermometer or a hygrometer.

Relative humidity markedly affects the ability of insects and other pests to survive in a particular environment, e.g. *Blattella germanica, Hofmannophila pseudospretella,* and many species of mites.

Remedial treatment. (*proc*) See IN-SITU TREATMENT.

Repellency test. (*proc*) See TEST METHODS.

Repellent. (*chem*) A substance which keeps away noxious insects and other pests; with little vapour action at a distance, hence usually requiring direct contact. Principally synthetic compounds, usually odourless, of limited life, applied to the skin or clothing (against biting flies) and sometimes to packaging materials (against cockroaches). Examples are DIETHYL TOLUAMIDE, DIMETHYL PHTHALATE, ETHOHEXADIOL, MGK REPELLENTS and PYRETHRUM.

Residual film. (*chem*) See RESIDUAL INSECTICIDE.

Residual insecticide. (*chem*) An insecticide applied in such a way that it remains active for a considerable period of time. Thus dusts, sprays, lacquers and baits may all be so described, if they contain a long-lasting insecticide with which an insect may make contact. The term is usually reserved, however, for insecticides applied by spray, which form a deposit on treated surfaces, remaining lethal for long periods.

Resistance (to pesticides). (*zoo*) The ability of a pest to tolerate a specified amount of a pesticide and survive. Not all individuals in a population may be killed by a pesticide, even though all may have picked up or acquired the same amount. The few that survive have a greater natural tolerance and form the breeding stock of the next generation. The progeny, having come from tolerant parents, survive the next application of pesticide, until, over a number of generations most of the individuals fail to succumb to the concentration used.

The characteristic for resistance is inherited from generation to generation, the pesticide acting continuously as a selecting agent. Resistance develops most quickly in pests (e.g. flies), which reproduce rapidly. It spreads most widely among pests which travel or are carried readily from one location to another. (See also CROSS-RESISTANCE.)

BAIT SHYNESS in rodents after eating a sub-lethal dose of rodenticide is a form of 'behavioural resistance'. This is an acquired characteristic of the individual and is not genetically transmitted to offspring.

Resmethrin. (*chem*) A synthetic pyrethroid (= NRDC 104), consisting of a mixture of isomers, first described in 1967. A white waxy solid. One of these isomers is Bioresmethrin (= NRDC 107), a liquid. All isomers are insoluble in water, but soluble in most organic solvents. Both resmethrin and bioresmethrin are more stable than pyrethrins, but decompose fairly rapidly on exposure to air and light. Both pyrethroids are good contact insecticides against a range of insects, and are many times more active than pyrethrins against house flies, but neither pyrethroid is so effectively synergised as pyrethrins, bioresmethrin hardly at all.

The acute oral LD_{50} (rat) of resmethrin is about 1500 mg/kg and of bioresmethrin 8000 mg/kg. Formulations replace pyrethrins in aerosols and space sprays; also as mixtures with pyrethrins and other synthetic pyrethroids to optimise knockdown and kill.

Respirator. (*equip*) See GAS MASK.

Resuscitation. (*proc*) See ARTIFICIAL RESPIRATION.

Returned goods. (*manuf*) Products returned to a manufacturer through some defect in quality. Such goods should be kept in isolation on return to the factory since some may come from sources where pest control is either not practised or inadequate, thus offering the chance of infestation being introduced.

Returned crates used for the distribution of beverages, provide a ready source of reinfestation of bottling plants and breweries. Laundry baskets provide a means of carrying cockroaches into hotels and hospitals.

Rhipicephalus sanguineus. (*zoo*) Acari: Ixodidae. The Brown dog tick. Found throughout the U.S. causing blood loss and discomfort to dogs; brought into the home on pets from other infested properties, rarely from outdoors and not from direct contact with other dogs. A moult always occurs between change of hosts. In the U.S. the Brown dog tick feeds almost exclusively on dogs, rarely attacking man.

The adults are flat (3 mm long) red-brown with tiny pits over the back. When engorged after a blood meal, the enlarged body of the female (12 mm long) becomes grey-blue.

Eggs are laid in crevices, under plaster and carpets and may not hatch for several months under cool dry conditions. Nymphal development is complicated by the need for blood meals, periods spent off the host (when moulting occurs), and extended survival without food. A blood meal is essential before mating, the females becoming engorged with blood before egg-laying.

Resistance by the Brown dog tick to organochlorine insecticides occurs in many areas of the U.S. Organophosphorus insecticides are effective, especially CHLORPYRIFOS.

Rhizopertha dominica. (*ent*) Coleoptera: Bostrychidae. The Lesser grain borer. A pest of cereals, especially in warm climates (India, Australia and parts of the U.S.A.), uncommon in the U.K., except on imports. Unable to survive U.K. winters.

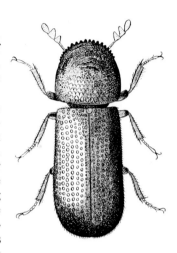

The adult beetle (2·5–3 mm long) is brown-black, cylindrical with a hump-like pronotum concealing the head from above. The elytra are strongly pitted. Capable of flight. Eggs are laid loose or attached to grains. Larvae bore into the grains, feeding within, and reducing them to husks. Pupae are formed inside the grains. The minimum period of development from egg to adult is 25 days (at 34°C) extending to 130 days under less favourable conditions.

Rhothane. (*p. prod*) TDE. An analogue of DDT with similar chemical properties, but of lower toxicity to man; introduced as a proprietary product in 1945 by Rohm & Haas Co. The technical product is a colourless crystalline solid, a mixture of isomers and related compounds, insoluble in water, but soluble in many organic solvents. Odourless, tasteless and non-flammable. Its insecticidal properties include stomach and contact activity; used as a mosquito larvicide.

The acute oral LD_{50} (rat) is 3400 mg/kg; chronic oral toxicity and dermal absorption are 3–4 times lower than DDT. Formulations include wettable powder, emulsion concentrate and dust.

Rice weevil. (*ent*) See SITOPHILUS ORYZAE.

Rickettsia. (*dis*) The generic name for a group of viruses which flourish in the alimentary tracts of ticks and lice causing various diseases in man (e.g. *Murine typhus*).

Roaches. (*ent*) See BLATTARIA.

Rodent control. (*proc*) The application of rodenticidal formulations to bring about the elimination of commensal rodents in and around buildings, in sewers and other locations where rats and mice constitute a pest problem. Also the application of PREVENTIVE PEST CONTROL measures to stop rodent entry and a recurrence of the problem.

There are eight good reasons for rodent control:

1) Monetary loss: there are no reliable estimates, but guesses suggest that the annual cost of rats and mice, in Britain alone, is £10–15 million.
2) Prevention of damage: almost every type of commodity, whether in production, store or use, is subject to rodent attack.
3) Contamination: in addition to food consumed, a far greater amount is fouled and made unfit for human consumption.
4) Loss of goodwill: damaged and fouled articles are unacceptable in trade.
5) Disease potential: rodents carry a number of diseases injurious to man.
6) Legislation: in certain countries, it is an offence to harbour rodents.
7) Distress: to many people rats and mice cause fear.
8) Working conditions: no one likes to be employed where rats and mice are present.

Control methods involve, primarily, poison BAITS, CONTACT DUSTS and FUMIGANTS (*q.v.*).

Rodentia. (*zoo*) Rodents: rats and mice (but also, squirrels, beavers, porcupines and coypus). The Order of mammals which have a single pair of incisor teeth in both the upper and lower jaws (Fig. 30). These teeth continue to grow throughout life and must be used continuously to keep their length constant. Hence the gnawing activity associated with these animals. The group comprises many genera and species among which *Rattus norvegicus*, *Rattus rattus* and *Mus musculus* are the only major indoor pests (but see also APODEMUS SYLVATICUS and VOLES). Under favourable conditions, rats and mice are capable of breeding throughout the year; a high reproductive ability results in rapid increases in rodent populations (see Figs. 32 and 33).

Rodenticide. (*chem*) A chemical substance which kills rodents, usually incorporated into a BAIT or CONTACT DUST (*q.v.*). Used principally for the control of *Rattus norvegicus*, *R. rattus* and *Mus musculus*. In baits, chemicals are used in two ways: 1) as CHRONIC POISONS (*q.v.*); anticoagulants of the hydroxycoumarin type (e.g. WARFARIN *q.v.*), or indane-dione type (e.g. CHLOROPHACINONE *q.v.*) and 2) as ACUTE POISONS (*q.v.*); varying in action, e.g. ZINC PHOSPHIDE, THALLIUM SULPHATE and ALPHACHLORALOSE (*q.v.*). Fumigants are also used to kill rodents in burrows (see CYMAG) and to disinfest ships (see HYDROGEN CYANIDE).

Rodent repellent. (*chem*) See ROTRAN R.55.

Rodent smears. (*zoo*) Body grease left by rodents on surfaces over which they move. Rodents tend to use regular routes and smears often accumulate where there is close body contact (e.g. along beams, beneath joists). These marks become accentuated with the accumulation of dirt (Fig. 82).

Rogor. (*p. prod*) See DIMETHOATE.

Ronnel. (*chem*) See FENCHLORPHOS.

Roof void. (*bldg*) The space between a ceiling and the roofing structure: a favoured location for many pests (rodents, birds, bats, insects) because of dark undisturbed harbourage, and not infrequently the presence of nesting materials. Additionally it provides a primary site for insect attack of timber (see e.g. ANOBIUM, HYLOTRUPES).

Access to roof voids is essential for inspection and treatment, often requiring ACCESS PANELS (*q.v.*) to be cut and constructed for otherwise closed areas. The user of pesticidal fluids should recognise the existence of electrical cables, water tanks, the flammability of the formulations used and the possibility of ceiling staining.

Rotenone. (*chem*) Derris. An insecticidal extract of the roots of the leguminous plants *Derris elliptica* and *Lonchocarpus* spp; first isolated in 1895. The crystallised extract is colourless, odourless, slightly soluble in water; soluble in many organic solvents. Of short persistence and used principally for the control of garden pests, ticks, lice and fleas.

Of low toxicity to mammals: the acute oral LD_{50} (rat) is 1500 mg/kg. Irritating to the eyes and mucous membranes. Most often used as a dust.

Rotran R.55. (*p. prod*) A rodent repellent of Phillips Petroleum: a waxy crystalline solid, pale brown with a sulphur-like odour, soluble in aromatic solvents and most petroleum oils, virtually insoluble in water. Promoted initially for the protection of buried seismograph cables against pocket gopher attack, with claims for repellency against commensal rodents. Oral contact is essential for R.55 to be effective; procedures have been developed for its inclusion in the insulation of telephone, geophysical and other types of cable; potential also for the protection of plumbing components of plastic composition.

Acute oral LD_{50} (rat) 1500 mg/kg. Formulations available include 25% emulsion for the treatment of soil in which cables are buried.

Rozol. (*p. prod*) See CHLOROPHACINONE.

Rust-red flour beetle. (*ent*) See TRIBOLIUM CASTANEUM.

Rust-red grain beetle. (*ent*) See CRYPTOLESTES FERRUGINEUS.

Sachet. (*chem, equip*) A 'unit' pack containing a measured dose of a pesticide, e.g. a rodenticidal bait, often used for insertion into stacks of foodstuffs, or as 'throw packs' to get them into inaccessible locations. Directions for use vary: in some instances the sachets are used closed, the rodent gnawing the paper to gain access; for mouse control, sachets are more effective if opened before use.

Salmonella. (*dis*) A major food contaminant. A genus of bacteria, pathogenic to man, causing food poisoning (salmonellosis) and other enteric disorders (gastroenteritis); transmitted by rodents and cockroaches through fouled food and infected droppings. Examples are *Salmonella analis*, *S. enteritidis*, *S. morbificans*, *S. oranienburg*, *S. typhimurium*. See also BIOLOGICAL CONTROL.

Sand snake. (*equip*) A length of canvas tubing filled with sand, used to anchor fumigation sheets at ground level to prevent the escape of gas. Discarded fire-fighting hose is suitable for the purpose. See also COAMING.

Sanitation. (*dis*) Measures for the promotion of health, especially, drainage, sewage disposal and the creation of conditions favourable for a pest-free environment.

Santobrite. (*p. prod*) See SODIUM PENTACHLOROPHENATE.

Saw-toothed grain beetle. (*ent*) See ORYZAEPHILUS SURINAMENSIS.

Scavenger. (*zoo*) An organism that feeds on dead organisms or on the wastes of others, including man, e.g. rats on food spillage, gulls on refuse tips and spider beetles on residues of foodstuffs in the cracks of floor boards.

Scillirocide. (*chem*) See RED SQUILL.

Sclerite. (*ent*) A hardened cuticular plate linked to others by intersegmental membranes, or fused with adjacent plates making up the outer layer of the insect integument. See STERNITE and TERGITE.

Sealing-up. (*proc*) The process of ensuring that a building, vehicle, or space containing a commodity under fumigation is gas-tight. Joins in fumigation sheets are sealed with adhesive tape (indoors), or rolled and clamped (outdoors, Fig. 37); SAND SNAKES and chains (Fig. 36) are used round the base. Buildings, trucks and other vehicles are often fumigated without sheeting, although some form of sealing around doors and other openings helps to make them more gas-tight. Useful sealing materials are gummed kraft paper, adhesive plastic tape and sponge rubber (around sliding doors of railway vehicles). Vehicles and trucks which are moved while under gas should have wire passed through the door hasps or locks making entry impossible. With all types of fumigation, appropriate warning signs should be posted.

Secondary poisoning. (*tox*) The accidental killing of an animal by it feeding on another which has already succumbed to direct, or primary, poisoning by a rodenticide or insecticide: as for instance a dog which eats a poisoned rat. Secondary poisoning occurs rarely; the risk is greatest with the use of 1) acute rodenticides of high toxicity (e.g. fluoracetamide, sodium monofluoroacetate, thallium sulphate), 2) compounds not readily detoxified in the body and 3) poisons to which domestic animals are especially susceptible.

A classic example occurred in the U.K. in 1964. In error, or contrary to label recommendations, fluoracetamide was used to control rats on a rubbish tip in Merthyr Tydfil, Wales. A pony consumed the bait and was subsequently sold as pet food: 40 dogs died in that locality in the next few days.

Secrete. (*zoo*) To hide or conceal, as cockroaches, in a harbourage. To produce by way of secretion, as certain body fluids.

Segregation. (*manuf*) The practice in food manufacturing of keeping incoming raw materials apart from processing areas; in warehousing, of keeping different classes of goods well separated from each other in storage. Goods known to be infested or susceptible to attack should be held apart from those which are not, thus helping to limit the spread of insect pests.

In food manufacture all raw materials, packaging materials and finished products should be geographically separated. This helps reduce infestation in processing areas (especially if raw materials are first fumigated) thus tending to keep insect infestation in finished products to a minimum. See also STOCK ROTATION.

Selectivity. (*chem*) The property of a pesticide of being more toxic to one animal (the pest) than to another (e.g. honey bee); ideally presenting no hazard to other animals and man. See NORBORMIDE and TARGET SPECIES.

Serviceman. (*name*) Operator. A person who carries out pest control. One who actually does the job, whether self-employed, an employee of a health department (local authority) or contractor. His aim is 1) to eliminate a pest, 2) if this is not possible to so reduce its numbers as to minimise the problem, and 3) to prevent its recurrence.

A serviceman is a key person in the Pest Control Industry: he should know where pests occur, the damage they do and how to control pests without introducing health risks. The service he gives is important to the health and comfort of the individual and the community (Fig. 69).

Sesamex. (*chem*) A synergist for pyrethrins and synthetic pyrethroids introduced in 1956 by Shulton Inc. under the trade name 'Sesoxane' and with some insecticidal activity. More costly than piperonyl butoxide and unstable. A straw-coloured liquid with faint odour. Acute oral LD_{50} (rat) is 2000 mg/kg.

A pyrethrum synergist with a similar name, but chemically unrelated, and non-insecticidal is Sesamin, a crystalline fraction of sesame oil, derived from the seeds of *Sesamum indicum*.

Sesamin. (*chem*) See SESAMEX.

Sesoxane. (*p. prod*) See SESAMEX.

Sevin. (*p. prod*) See CARBARYL.

Sewer. (*bldg*) A pipe or conduit designed to carry to a suitable outfall, waste products that are capable of being removed by the aid of water. Used for the drainage of buildings, and yards appurtenant to buildings, with the exception of those, as defined under DRAIN, for the removal of waste from buildings or yards within the same curtilage. A sewer is therefore communal in character, being a pipe or conduit into which the drains of two or more individual properties discharge.

Sewers are of two kinds, viz. 'Private sewers' and 'Public sewers': the latter is a sewer which is, or becomes, vested in a local authority, in as much as rodent control in those sewers also becomes a local authority responsibility.

Sewer rat. (*zoo*) See RATTUS NORVEGICUS.

Sewer treatment. (*proc*) The lifting of manhole covers and placement of bait on BENCHINGS (*q.v.*) for the control of rats (*Rattus norvegicus*) which infest and shelter in sewers (Fig. 83). Elimination of the sewer population can bring about a marked reduction in the number of rats living above ground: of vital importance in modern cities and the first step in any campaign against rats.

Treatment often involves lifting gear to remove the manhole covers, the use of fluoracetamide (2% baits), and a means of placing the bait with a telescopic tube or loose bottomed box. Where there is no benching, baits are suspended above the effluent in a muslin bag.

Shad roach. (*ent*) See BLATTA ORIENTALIS.

Sheep tick. (*zoo*) See IXODES RICINUS.

Shelter tube. (*ent*) Mud tube. The cover of soil particles, saliva and excreta made by SUBTERRANEAN TERMITES over their trails on brickwork, concrete and plaster, providing concealment and protection from desiccation during movement to and from soil and food source (Fig. 85).

Shigella. (*dis*) A genus of bacteria, containing species pathogenic to man. For example, *Shigella alkalescens* (dysentry) and *S. paradysenteriae* (summer diarrhoea in children), both of which have been found in the intestinal tracts of cockroaches.

Ship rat. (*zoo*) See RATTUS RATTUS.

Shoxin. (*p. prod*) See NORBORMIDE.

Silica aerogel. (*chem*) See AEROGEL.

Silo. (*bldg*) A bulk storage unit for raw foodstuffs (e.g. grain, groundnuts, pulses) circular or square in section, constructed from metal, wood or concrete, with a valve at the base for discharge. Silos invariably provide crevices for resident populations of pest insects and require regular cleaning and spraying of internal surfaces to prevent CROSS-INFESTATION of products. They are convenient containers for fumigation; some are equipped with circulatory fumigation plant. See also HOPPER.

Silverfish. (*ent*) See LEPISMA SACCHARINA.

Siphonaptera. (*ent*) The Order of insects containing the fleas; small wingless insects flattened side to side, red-brown with backwardly directed spines and well-developed legs for jumping. All fleas when adult are parasitic on warm-blooded animals. Larval stages feed in the nest of the host on fragments of skin, feather, dried blood and excrement. Examples are *Pulex irritans, Ctenocephalides felis* and *Ceratophyllus gallinae*.

Site pre-treatment. (*proc*) See TERMITE CONTROL.

Sitona. (*ent*) Coleoptera: Curculionidae. A genus of plant weevils, often brightly coloured, feeding mainly on clover, grasses and sometimes garden plants, occasionally getting into the home, but not causing damage indoors. Often harvested with cereals, being about the same size as wheat grains.

Sitophilus. (*ent*) Coleoptera: Curculionidae. Previously *Calandra*. Pests of grain on farms, in mills, warehouses and at major grain storage centres. Especially prevalent in cereals in developing countries. Weevils (2–4 mm long), with a hard cylindrical body, reddish brown to black with a long snout or rostrum and elbowed antennae ending in a distinct club. The larvae, white, legless, with a pale brown head living wholly within the grain. Three species are recognised:

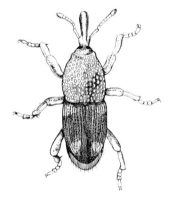

Sitophilus granarius (Fig. 92): the Grain weevil, or Granary weevil. Punctures on the thorax distinctly oval; the lines of punctures on the elytra widely separate. The hind wings atrophied; unable to fly. Survives U.K. winters. Development from egg to adult takes 4 weeks (27°C); the adult lives for about 3 months, or longer at lower temperatures. Breeding stops below 13°C.

Sitophilus oryzae: the Rice weevil; common on wheat and rice. Punctures on the thorax round; elytra with two pairs of orange spots and the lines of punctures narrowly separate. Capable of flight at high temperatures. Succumbs to U.K. winters. Development from egg to adult takes 4 weeks (27°C). (Illustrated above.)

Sitophilus zeamais: the Maize weevil. Very similar in appearance to *S. oryzae*, separated by detailed examination of the genitalia. Capable of flight at high temperatures. Speed of development slightly more rapid than *S. oryzae*.

Sitotroga cerealella. (*ent*) Lepidoptera: Gelechiidae. Angoumois grain moth. Named after Angoumois in central France where it was first noticed as a serious pest. Of minor importance in the U.K., occasionally imported on shipments of cereal grains. Overseas, infestation may arise in grain in the field.

A small moth (9 mm long) pale yellow brown, the wings with wide fringes; the hind wings have a finger-like extension. Eggs are laid on the grain soon after harvesting; larval development occurs within individual grains, several in one seed. Adults do not feed. Development from egg to adult takes 6 weeks (21–26°C).

Skirting board. (*bldg*) Base board (and other local names). The length of timber along the base of a wall protecting it from damage. Frequently with gaps between it and the floor and the wall, often providing a harbourage for cockroaches and other domestic insect pests.

Slender-horned flour beetle. (*ent*) See GNATHOCERUS MAXILLOSUS.

Slow release strip. (*chem*) A formulation of dichlorvos (20%) in a plastic base, designed to give extended and slow release of dichlorvos vapour into the air, for indoor fly control, and in a smaller size for the control of textile pests in domestic cupboards and wardrobes. Also as 'dog collars' for the control of animal parasites.

There are many variants of the product for fly control, including dichlorvos absorbed into inert granules, the rate of release controlled by regulating vaporisation from the container. Permitted use varies in different countries where food is exposed and where people are subject to prolonged exposure (e.g. hospital wards).

Slow release strips have largely replaced other automatic insecticide dispensers such as THERMAL VAPORISING UNITS (*q.v.*).

Smoke generator. (*chem*) A finely ground insecticide incorporated into a pyrotechnic mixture which, when ignited, gives off a cloud of insecticidal particles spreading over a wide area, leaving a deposit of insecticide mainly on horizontal surfaces.

A smoke generator is the easiest way of achieving a wide distribution of insecticide and sometimes of getting an insecticide into inaccessible locations. Account should be taken of loss of smoke by ventilation from spaces such as roof voids. An insecticidal smoke is not a fumigant; the insecticide is particulate and not a gas. Smoke generators most commonly incorporate lindane.

Smoky-brown cockroach. (*ent*) See PERIPLANETA FULIGINOSA.

Sodium arsenite. (*chem*) A mixture of sodium ortho-, meta- and pyro-arsenites. Once used in baits for the control of ants (3%), cockroaches, and for the mothproofing of carpets (0·5%). Also as a soil poison in termite control, in rodenticidal baits (for mice), and as a non-selective weedkiller. All these uses have now been superseded by less toxic chemicals.

Sodium arsenite is a white-grey powder. Freely soluble in water; hygroscopic and oxidised in contact with air; containers must be air-

Fig. 81 Action of a tactile bird repellent applied to a building ledge: Upper picture; pigeon attempting to alight. Middle; bird in contact with repellent. Lower; pigeon repelled. (*Sunday Telegraph*)

tight. An exceedingly potent compound to man, domestic animals and birds. Acute oral LD_{50} (rat) 10–50 mg/kg.

Sodium fluoride. (*chem*) An inorganic compound once used as a stomach poison for cockroach control, as a dust (25–75%), either alone, or admixed with pyrethrum, or as poison baits (5–10%). A dense, colourless crystalline solid; odourless, of low volatility and moderately soluble in water. Its use in pest control is no longer justified; alternative, less toxic and more effective insecticidal dusts and baits are available. Phytotoxic; used also as an ingredient of timber preservatives.

The lethal dose to man is 75–150 mg/kg. The lowest reported fatal dose is 2 g. Because of the hazard of mistaking sodium fluoride for domestic food products, the need for a colouring agent is strongly emphasised.

Sodium monofluoroacetate. (*chem*) Sodium fluoroacetate, Compound 1080. Used as an acute rodenticide in liquid baits. Application in the U.K. is restricted to the DERATTING OF SHIPS (*q.v.*) and drains in port areas (see POISONS RULES), according to a well-defined safety procedure.

The technical material is colourless, odourless, very soluble in water and hygroscopic when exposed to air. W.H.O. standards apply to purity of the commercially available product, which contains 0·5% nigrosine dye.

Intensely poisonous to man and other vertebrates. Acute oral LD_{50} (rat) is 1·5 mg/kg; rapidly absorbed through the intestinal tract, absorbed also through cuts and abrasions of the skin.

Formulations available include a 0·5% solution for dilution to 0·25%. Safety is best ensured by storage in a POISONS CUPBOARD (*q.v.*), restriction of use to experienced personnel, and emphasis on handling with extreme care.

Sodium pentachlorophenate. (*chem*) A salt of pentachlorophenol known under the Monsanto trade name, 'Santobrite'; buff flakes, soluble in water, insoluble in petroleum oils. Used as a fungicide where a water formulation is required. Cf. PENTACHLOROPHENOL.

Soil poisoning. (*proc*) See TERMITE CONTROL.

Soil treatment. (*proc*) See TERMITE CONTROL.

Soldier termite. (*ent*) A caste of termites. Sterile, with the specific function of protecting a colony from enemies. The head is enlarged, often heavily chitinised, with well-developed mandibles (Fig. 14). In some species, nasutes are produced: soldiers in which the front of the head is formed into a rostrum, capable of ejecting a fine stream of toxic fluid from a frontal gland at an enemy. Nasutes are usually smaller in size and the mandibles much reduced.

Solenopsis. (*ent*) Hymenoptera: Formicidae. Fire ants. A genus of small ants (workers 1·5–6 mm long) yellow-brown to black, the pedicel with two knobs. Pests of warm climates; absent from Britain. Objectionable by their presence and the severe reaction to their sting. Nests occur outdoors in dry soil, usually with small craters on the surface, sometimes

Fig. 82 (*above*) Body smears of *Mus musculus* on a wall, the mice running over the top and down the side of boxes, some of which have since been removed.

Fig. 83 (*below*) Treatment of a sewer for rat control: introduction of bait onto the benching by means of a telescopic plastic tube.

Fig. 84 (*above*) Mounds of subterranean termites in a sub-floor area. (Shell).
Fig. 85 (*below*) Shelter tubes of subterranean termites on composition blocks providing cover beneath which the insects move from the soil to the floor above. (Velsicol).

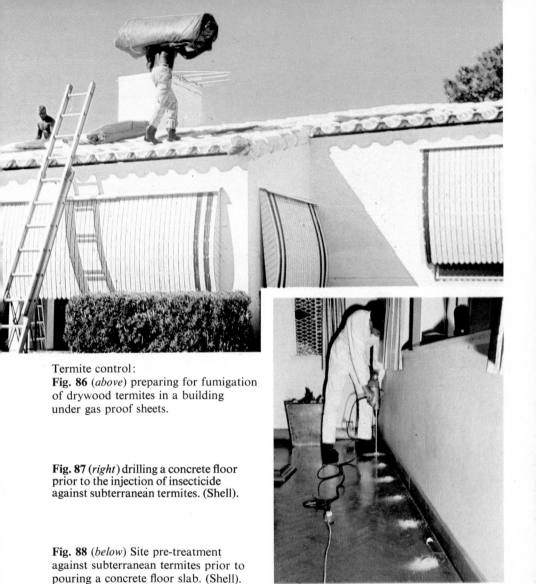

Termite control:

Fig. 86 (*above*) preparing for fumigation of drywood termites in a building under gas proof sheets.

Fig. 87 (*right*) drilling a concrete floor prior to the injection of insecticide against subterranean termites. (Shell).

Fig. 88 (*below*) Site pre-treatment against subterranean termites prior to pouring a concrete floor slab. (Shell).

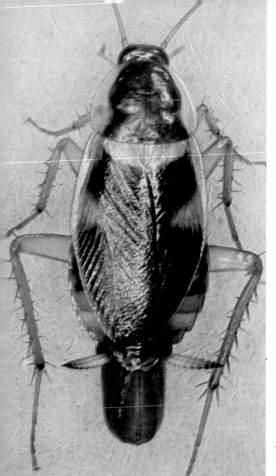

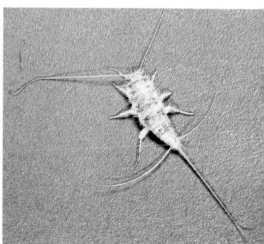

Fig. 89 (*left*) *Supella longipalpa* (female) with oötheca. The brown brands of the adults are not so well defined as in nymphs.

Fig. 90 (*below*) *Thermobia domestica* showing the three long tail bristles.

Fig. 91 (*right*) Treatment of a refuse area of an apartment block, using power spraying equipment. In warm climates frequent collection of domestic refuse is necessary to discourage rodents, cockroaches, and flies.

Fig. 92 (*left*) *Sitophilus granarius* on wheat.

Fig. 93 (*centre*) *Tribolium castaneum*, another stored product pest.

Fig. 94 (*bottom*) Evaluation of insecticides by topical application to *Blattella germanica*.

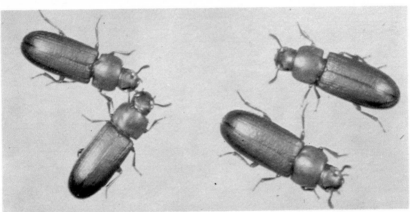

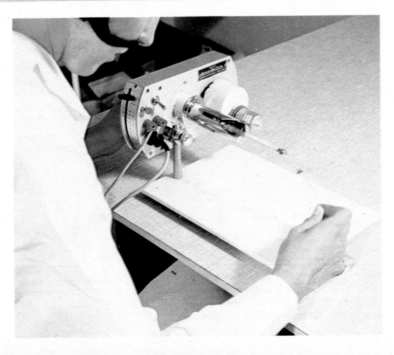

Fig. 95 Evaluation of insecticides by the contact of test cockroaches with treated surfaces. Glass is used in this instance because it is non-absorbent. The test insects are confined under inverted funnels and transferred onto the treated surfaces for different exposure periods.

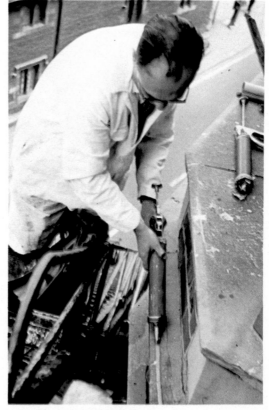

Fig. 96 (*above*) Control of feral pigeons by live trapping, a technique for reducing populations in a city area: the birds have been encouraged into the trap by pre-baiting and are then killed humanely.

Fig. 97 (*right*) Application of a tactile repellent by caulking gun to building ledges to deter feral pigeons and starlings. Access is obtained by turn-table ladder.

Vespula vulgaris
Fig. 98 (*above*) Worker taking the nectar of ivy flowers;
Fig. 99 (*below*) Nest hanging from the roof of an outbuilding.

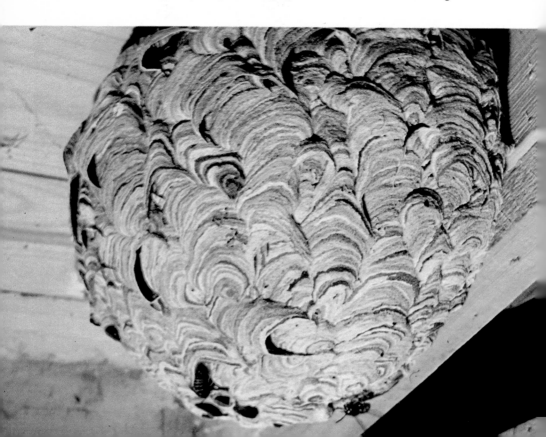

beneath stones and concrete, occasionally under and in the walls of properties. Fire ants travel in columns and enter properties in search of sweet foods, fats, nuts, cereals and meats; they also take the hone/dew of aphids.

Pest species include the Fire ant (*S. xyloni*), the Tropical fire ant (*S. geminata*), and the Imported fire ant (*S. saevissima*)—an agricultural pest with a severe sting. The genus also includes the smaller yellow Thief ant (*S. molesta*) which likewise invades kitchens and larders while foraging for food.

Control measures may involve treatment indoors (baits) and outdoors by soil treatment (e.g. with aldrin).

Solubility. (*chem*) In pest control, the property of a solid to dissolve in a liquid. Equally, the property of a SOLVENT (*q.v.*) to dissolve added solids. The extent to which a solid dissolves depends on its own characteristics, as well as those of the solvent and the temperature.

Solute. (*chem*) The substance dissolved in a solvent (e.g. the insecticide in an oil spray).

Solution. (*chem*) A solvent containing dissolved solids (e.g. odourless kerosene and lindane). Also a mixture of miscible liquids.

Solvent. (*chem*) A liquid in which a chemical may be dissolved. In some formulations (most oil sprays) the solvent dissolves the insecticide and acts as the DILUENT. When sprays are applied to surfaces, the solvent evaporates leaving a coating of insecticide. The rate of evaporation depends on the solvent used, e.g. kerosene dries slowly, xylene rapidly.

Sowbugs. (*zoo*) See ISOPODA.

Space treatment. (*proc*) The treatment of air by insecticidal mists or fogs, usually of synergised pyrethrins or dichlorvos, for rapid knockdown and kill of flying insects, e.g. flies, moths of stored foodstuffs. Occasionally also for the control of crawling insects, e.g. *Lasioderma serricorne* by the 'fallout' of droplets onto surfaces. Equipment used depends on the volume to be treated, e.g. pressurised aerosols, mechanical and thermal FOGGING MACHINES (*q.v.*).

Space treatment is temporary; it may help to prevent the spread of an infestation, reduce egg-laying (e.g. on finished products) but does not eliminate an infestation. Repeated applications are necessary for the technique to be of value. Space spraying should be carried out in conjunction with the treatment of breeding sites, using residual sprays, supported by improved sanitation. Of no merit for cockroach control. See also SLOW RELEASE STRIP, THERMAL VAPORISING UNIT and NON-STOP AEROSOL.

Sparrow, English or **House.** (*zoo*) See PASSER DOMESTICUS.

Specific gravity. (*phy*) Of a substance: the ratio of its weight to that of the same volume of water, at the same temperature and pressure. Information on specific gravity is often necessary in formulating oil-based sprays when the concentration of active ingredient is expressed on a

WEIGHT/WEIGHT basis. The specific gravity of a liquid is measured with a hydrometer.

Specificity. (*chem*) The property of a pesticide in controlling the TARGET SPECIES (*q.v.*) without effect on other animals and man.

Spider beetles. (*ent*) See PTINIDAE.

Spiders. (*zoo*) See ARANEAE.

Spillage. (*manuf*) Wasted food accumulating as residues during handling and manufacture, in loading and unloading areas of factories, rail sidings, despatch bays; material lost from conveyors, chutes and ducts in a manufacturing process, accumulating between floor boards and beneath machinery, resulting in the formation of conditions favourable for pest infestation. See PREVENTIVE PEST CONTROL.

Spiracle. (*ent*) An opening on the thorax or abdomen of insects which permits entry of air into the TRACHEAE (*q.v.*). The portal of entry of insecticidal fumigants.

Spirochaete. (*dis*) A genus of bacteria with a wavy outline, hence the name which means 'coiled hair', causing disease in man and animals, e.g. Weil's disease. See LEPTOSPIROSIS.

Spot treatment. (*proc*) Application of an insecticide to a small area of a building, or to separate processing machines, where a localised infestation exists. Often the means of checking insect infestations in flour mills between annual fumigations.

Spray. (*phy*) A collection of numerous liquid DROPLETS (*q.v.*), usually of an insecticide in water or oil, formed by pressurising the fluid through a fine jet. The size of the droplets (100–1000 μ) and spray pattern ('flat fan', 'pin stream', 'solid cone', 'hollow cone') depend much on the size and shape of the orifice (nozzle) and the pressure applied. The form in which insecticides are most often applied, as wettable powders, emulsions and in oil solution.

Sprayers. (*equip*) The equipment most widely used in pest control, varying in design and capacity for the pest problem and location.

1) *Hand sprayer:*

A portable hand-operated sprayer, used mainly for indoor work, consisting of a canister of capacity up to 10 litres and a hand-operated pump (plunger) which pressurises the air above the fluid level. The pressurised air forces the liquid out via a short tube terminating in a nozzle. For some work a length of flexible hose and a LANCE (*q.v.*) may be interposed between the canister and nozzle. The hosing should withstand the chemical action of water and the solvents in oil-based sprays. Hand-sprayers should be thoroughly cleaned daily and maintained in perfect working order.

2) *Powered sprayer:*

A sprayer pressurised by electrically powered or petrol-driven compressor. Capacity varies from 50–500 litres; often mounted on a vehicle, with the compressor operated by the drive of the vehicle.

Also electrically-driven centrifugal pumps used principally for IN-SITU TREATMENT of wood-boring insects, termite control and outdoor spraying of insecticides to refuse tips or other areas (Fig. 91). See also KNAPSACK SPRAYER.

Springtails. (*ent*) See COLLEMBOLA.

Stability. (*chem*) Of a chemical, its ability to remain unchanged under a variety of conditions; of pesticides, their ability to remain biologically active for long periods of time when applied; of insecticide emulsions, the ability of the oil droplets containing the dissolved active ingredient to be dispersed in water in the ready-to-use spray; of rodenticidal baits, the ability of the formulation to withstand loss in rodenticidal action, stay free from mould growth, and remain palatable in use.

Stack. (*manuf*) An orderly arrangement of bagged or crated commodities, one upon the other, optimising the use of floor space; built (preferably on PALLETS, Fig. 78) to allow inspection for rodent and insect infestation, and on a concrete floor or gasproof sheet if intended for fumigation. In this respect it is important that the fumigant should not be allowed to discharge directly into the commodity near the gas outlets.

Stadium. (*ent*) See INSTAR.

Stage. (*ent*) See INSTAR.

Staining. (*chem*) The discoloration of surfaces by pesticides; minimal with most oil-based sprays, but marks may occur at the edges of a sprayed area if dirty before treatment; water staining of fine fabrics can occur with emulsions; visible deposits may be objectionable on surfaces sprayed with wettable powders. Some cockroach baits (pastes and gels) mark wallpapers. Certain organophosphorus compounds (e.g. fenitrothion) cause brown staining in depth on alkaline surfaces (e.g. marble floors and wall decor). If in doubt examine the problem on a small test area first.

Staphylococci. (*dis*) A group of bacteria, some of which are pathogenic to man, typically associated with such infections as boils and abscesses. A number of Staphylococci (e.g. *Staphylococcus aureus*) have been found naturally infecting cockroaches, in their faeces, intestinal contents and on the outer surface of the insects.

Starling. (*zoo*) See STURNUS VULGARIS.

Steam fly. (*ent*) See BLATTELLA GERMANICA.

Stegobium paniceum. (*ent*) Coleoptera: Anobiidae. The Biscuit beetle or Drugstore beetle (in the U.S.A.). Infests many commodities (grain, flour, cocoa beans, spices); sometimes attacks leather goods, books and other manufactured products. A pest of larders, shops and packaged foods (e.g. breakfast cereals). Similar in appearance to LASIODERMA (*q.v.*) and ANOBIUM. See illustration on page 20 (*q.v.*).

Eggs are laid in crevices; the larvae (5 mm long) are off-white, C-shaped with numerous short colourless hairs; they can survive starvation for many weeks. The adult (2–3 mm long) is red-brown, densely covered with short yellow hairs and does not feed. The minimum period of development from egg to adult is 4 weeks (30°C); only one generation a year under cooler conditions.

Sterilant. (*chem*) See CHEMOSTERILANT.

Sterilisation. (*proc*) The disinfection of soils, e.g. with methyl bromide, killing bacteria, fungi, nematodes and the embryos of weed seeds. Also the disinfection of pharmaceutical products (sutures, syringes). The killing of bacteria in food products (e.g. spices) with ethylene oxide, and by high temperature: treatment above 121°C by moist heat in an autoclave, or by dry heat above 160°C for two hours. Pasteurisation is partial sterilisation, i.e. treatment at 60–70°C for 30 minutes which kills some bacteria but not all. See also CHEMOSTERILANT.

Sternite. (*ent*) A cuticular plate forming the ventral surface of each segment of an insect.

Sticky board. (*equip*) Glue board. A physical method of rodent control most often used to eliminate the occasional survivor of baiting treatments; more effective against mice than rats. A piece of hardboard or thick cardboard (30–60 cm square) is coated with 3 mm of tacky varnish or thixotropic gel designed to hold rodents which come in contact with it. Placed on rodent runs; their efficiency is much improved by baiting around the edge of the board. This technique may be permitted for bird control in some countries.

Stillage. (*equip*) A framework for keeping articles off the floor. An important feature of good storage in all premises, giving sufficient clearance for pest inspection. See also STORAGE. Useful in pesticide stores for supporting large drums allowing fluids to be dispensed into smaller containers with minimum spillage.

Stock rotation. (*proc*) In warehousing the practice of using the oldest stocks first, making a contribution to pest control by minimising the time during which pest populations can grow within stacks or individual containers.

Stomach poison. (*chem*) A chemical which must be swallowed to cause death (e.g. rodenticides and insecticides incorporated in baits). A term used to describe insecticides which kill more readily when ingested than by contact. Cf. CONTACT INSECTICIDE.

Storage. (*proc*) The accommodation of goods, best practised in a manner which reduces harbourage and therefore pest incidence. The important requirements for this in the food industry are:

1) all areas be accessible for cleaning and inspection,
2) damage to containers be reduced to a minimum, thereby cutting the amounts of spillage available, e.g. to rodents,
3) all goods be kept at least 50 cm from walls (Fig. 51), (see CRASH BARRIER),
4) sufficient gangways be left for inspection between stacks,
5) all goods be kept off floors on PALLETS (Fig. 78), STILLAGE, or storage shelves (but not DUCK BOARDS),
6) all areas be well ventilated,

7) the buildings used be in good repair and effectively proofed against pest entry.

Stored product pests. (*ent*) Specifically insects which attack edible food-stuffs in storage, transport and manufacture. In the wider sense, including also insect pests of larders, pantries, shops and catering establishments. Not usually insect pests of stored timber, leather, paper goods and textiles, nor rodents and birds, although these too are pests of stored products.

Strychnine. (*chem*) An alkaloid plant extract (*Strychnos nux vomica*), available also as water soluble strychnine sulphate. A very fast acting poison, once used for mouse and mole control; unacceptable to rats. Acute oral LD_{50} (rat) is 5 mg/kg. Legal restrictions on use are imposed in many countries: no longer justified as a pest control chemical because of its high toxicity, rapid action, and the existence of more effective alternatives.

Stupefying substance. (*chem*) See ALPHACHLORALOSE.

Sturnus vulgaris. (*zoo*) Starling, European starling in U.S. A pest bird in cities, roosting on the ledges of buildings at night, defacing and eroding the stonework by their droppings; their chatter becoming almost incessant and causing considerable annoyance when starlings congregate in large numbers.

Starlings are omnivorous, migratory, moving in large flocks: the British population swelled each year by migrants from N. Europe. They feed during the day some 15–30 km away from their roosting site, seeking the warmth of cities at night; usual nesting sites are in trees and the eaves of buildings. There are 5–7 eggs/clutch (Fig. 71), one brood each year, hatching time about 2 weeks and fledglings leave the nest after a further 3 weeks.

Control methods consist of bird proofing and the use of tactile repellents. See BIRD LAWS and BIRD CONTROL.

Subterranean termites. (*ent*) The most destructive pests of buildings in warm climates; these termites require a substantial supply of moisture for maintenance of the health of the colony, which is obtained from the soil. All are soil-nesting species, some building mounds above ground (Fig. 84), movements to and from sources of food within SHELTER TUBES (Fig. 85). Capable of penetrating narrow cracks in concrete. Trail following markedly developed.

Control methods involve treatment of soil before building construction, or after construction by drilling walls and floors followed by irrigation. (See TERMITE CONTROL.) Also by the use of pre-treated timber. Species of the genus *Reticulitermes* comprise the most important subterranean termites in the U.S.

Sucking lice. (*ent*) See ANOPLURA.

Sulfoxide. (*p. prod*) An organic compound with negligible insecticidal action of its own, but which acts as a potent SYNERGIST for pyrethrins and synthetic pyrethroids. The synergising action on pyrethrins is about five times higher than on allethrin.

The technical product is a brown liquid, insoluble in water, but soluble in many organic solvents and of low toxicity: acute oral LD_{50} (rat) about 2000 mg/kg. Formulations include an oil concentrate and emulsion concentrate.

Sulphaquinoxaline. (*chem*) A bactericidal compound incorporated into baits containing anticoagulant rodenticides, with claims for improved action. Sulphaquinoxaline is said to deplete the microflora of the gut of rodents, thus reducing internal production of vitamin K, an antidote to poisoning by anticoagulants, enhancing the action of the active ingredient.

Sulphuryl fluoride. (*chem*) An insecticidal fumigant introduced in 1957 by Dow Chemical Co., under the trade name 'Vikane'. A colourless, odourless gas; stable, non-corrosive, harmless to fabrics but phytotoxic. Three times heavier than air. Used for the fumigation of structures against drywood termites (Fig. 86). Not for use on foods destined for human or animal consumption. Acute exposures to man are about one third the toxicity of METHYL BROMIDE.

Sumithion. (*p. prod*) See FENITROTHION.

Supella longipalpa. (*ent*) Dictyoptera: Blattidae. The Brown-banded cockroach. Occurring over considerable areas of the tropics and subtropics of the Old World, much of Africa outside the forested areas of the West Coast, and especially abundant in the Sudan. First recorded in the U.S. (Miami) in 1903, but now established in heated homes and apartments throughout the United States, where it is a major domicillary pest, having increased rapidly in abundance during the last 40 years. It is known to occur in buildings in Europe, but at present there are only a few recorded occurrences of this cockroach in Britain. Unlike the German cockroach which is usually confined to kitchens and eating areas, the Brown-banded cockroach is an active species which spreads throughout buildings, preferring locations high up in heated rooms. It is found typically in desk drawers, bedding and bedroom furniture, behind pictures and picture moulding, and beneath wallpaper. Widespread dispersal in homes makes the Brown-banded cockroach especially difficult to control.

The adult is small (10–14·5 mm) similar in size to the German cockroach, the tegmina covering the abdomen in the male but rarely reaching the apex in the female. Coloration is variable, generally ochraceous buff, the pronotum dark but for a paler, transparent area along each lateral edge. Brown bands on the mesonotum and first abdominal segment of nymphs give this cockroach its name. Two chestnut-coloured, fairly irridescent areas, on the tegmina give the adult an attractive appearance (Fig. 89).

Among pest cockroaches, *Supella* produces the smallest oötheca, with the surface marked into segments (about 8–9) and containing about 16 eggs. It is deposited soon after formation and takes 5 weeks to hatch at 30°C; development to maturity takes about 3 months.

Super rat. (*name*) A term used by the press to describe a rat 'of almost supernatural powers, much larger and voracious than rats generally, existing in massive numbers and outwitting man in his attempt at control'; used in the U.K. especially to describe rats resistant to warfarin. The same applies to Super mouse.

Supplementary reproductive. (*ent*) A caste of termites, wingless or with very short wings; individuals are developed as needed. Fertilised females take the place of the queen, which may die, or provide additional queens within the same nest, or establish new nests nearby.

Supplier. (*manuf*) A trading organisation or individual. For the food manufacturing industry, those providing equipment, raw materials for processing, and the packaging for finished products.

All suppliers handling foodstuffs are vulnerable to infestation, which if not controlled can be passed on to users. As part of PREVENTIVE PEST CONTROL (*q.v.*) the subject of infestation should be discussed with suppliers, visited if possible by the consumer's hygiene officer, and assurances sought that satisfactory pest control measures are being employed. Equipment and packaging suppliers should be treated similarly.

Incoming raw materials should be inspected when received at the user's factory. Staff should be taught to look for obvious signs of rodent activity and insect damage. Any evidence should be reported and lines of communication established to ensure that the appropriate and agreed action is taken. See also TRANSPORT.

Surfactant. (*chem*) See WETTING AGENT.

Survey. (*proc*) See INSPECTION.

Susceptible. (*zoo*) Readily affected by a pesticide; capable of being killed by commonly used, safe or economic concentrations. Comparison of susceptible and resistant strains is often made in the laboratory to measure the performance of potential new pesticides. See RESISTANCE.

Suspension. (*phy*) Insoluble particles uniformly distributed in a fluid prevented from settling by agitation, viscosity, or some other physical condition: as in ready-to-use sprays containing a wettable powder. Good DISPERSION (*q.v.*) and suspension are necessary to ensure that a uniform application of insecticide is achieved.

Swallow fly. (*ent*) See CRATAERINA PALLIDA.

Swarming. (*ent*) The emergence from nests of large numbers of insects at certain times of the year, e.g. ants in late summer, when winged males and females take part in the nuptial flight. Also the swarming of winged reproductives of subterranean termites in spring and autumn leading to pairing and the formation of new colonies. The congregation of large numbers of flying insects in a small area, e.g. Chironomids over water, or *Thaumatomyia* in houses.

Swingfog. (*equip*) See FOGGING MACHINES.

Synergism. (*chem*) The improvement in effect of a pesticide when a chemical of negligible biological activity is mixed with it, allowing the use of the pesticide at a much lower concentration (and cost). Examples of synergists are piperonyl butoxide (the usual synergist for pyrethrins), sesame oil and sulfoxide. Cf. POTENTIATION, an increase in mammalian toxicity resulting from admixture.

Synthetic pyrethroids. (*chem*) Substances of similar chemical structure to natural pyrethrins, some with high KNOCKDOWN (*q.v.*) and low kill, others with little knockdown but good kill. Generally not synergised as effectively as pyrethrins.

The first pyrethroid produced on a large scale was ALLETHRIN having about 70% the activity of pyrethrum. More recent synthetics include BIOALLETHRIN, RESMETHRIN (= NRDC 104), bioresmethrin (=NRDC 107) and TETRAMETHRIN. There are additionally a number of NRDC compounds currently undergoing evaluation.

T

Tactile repellent. (*chem, equip*) A substance or structure which by physical contact causes an animal to move to another location, e.g. the thick gels, plastic compositions and various metal devices designed for repelling feral pigeons (Fig. 81). The effectiveness of gels depends much on care in application (Fig. 97), and the roosting pressure of the population. Wires and pointed metal strips, although more difficult to install, can be effective on limited areas of buildings, providing alternative roosts are available nearby. Electrically charged wires soon fall into disrepair and have a short life. All these methods rely for effectiveness on breaking the flock roosting habit. See ENTANGLEMENT.

Take. (*name*) A colloquial word indicating the *amount* of BAIT eaten or removed by a pest.

Target species. (*zoo*) The pest species to be controlled; the method and materials used not affecting other forms of life in the environment—the non-target species. This objective is more likely to be achieved by the use of specific pesticides than broad spectrum pesticides. See SPECIFICITY.

Tarpaulin. (*equip*) See GAS-PROOF SHEET.

Tarsus. (*ent*) the 'foot' of an insect, consisting of one to five segments, the last bearing the tarsal claw, often with a pad (arolium, or pulvillus in flies) between.

Tartar emetic. (*tox*) See ANTIMONY POTASSIUM TARTRATE and EMETIC.

Taxonomy. (*zoo*) The arranging of living organisms in a CLASSIFICATION (*q.v.*) based on their similarity and diversity of structure. A person so involved is a taxonomist.

TDE. (*chem*) See RHOTHANE.

Technical material. (*chem*) A chemical pesticide (usually of 90–99% purity) in its undiluted form, as supplied by a manufacturer. Sometimes available with less odorous properties (e.g. organophosphorus compounds), often referred to as 'Premium grade'. Used to make CONCENTRATES (*q.v.*) from which ready-to-use formulations are prepared.

Tegenaria. (*zoo*) Arachnida: Araneae. House spiders. Commonly occurring

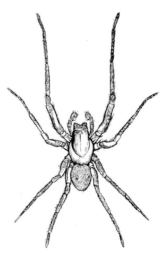

spiders in homes; harmless, fast runners, objectionable only by their presence; long-lived. Spiders with long legs, nearly twice as long as the body, producing little by way of a web. Eggs laid in a silken sac. Examples are *T. domesticus* and *T. atrica*, both carnivorous, preying mainly on insects.

Tenate (1080). (*chem*) See SODIUM MONOFLUOROACETATE.

Tenate-one (1081). (*chem*) See FLUORACETAMIDE.

Tenebrio. (*ent*) Coleoptera: Tenebrionidae.

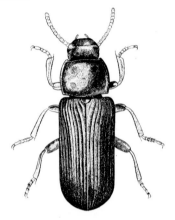

Mealworm beetles. Large beetles (15 mm long), the adults and larvae occurring as minor pests of cereals, seldom in large numbers. They are scavengers in dark, damp locations (warehouses, mills, larders and basements) where residues accumulate. Often breeding in birds' nests in roof spaces. Two species are pests of foods; *T. molitor*, Yellow mealworm beetle and *T. obscurus*, the Dark mealworm beetle.

Adults are robust, dark brown to black. Larvae (28 mm long) are yellow (in *T. molitor*) or white tinged with brown (in *T. obscurus*); typical segmented appearance, omnivorous, resistant to starvation. Development from egg to adult takes 9–15 months, or longer under adverse conditions. Adults live for 2–3 months.

Tenebrionidae. (*ent*) The family of the Coleoptera containing numerous pest beetles of stored foods, mostly grain and cereal products. See ALPHITOBIUS, GNATHOCERUS, TENEBRIO and TRIBOLIUM. See also BLAPS.

Tenebroides mauritanicus. (*ent*) Coleoptera:

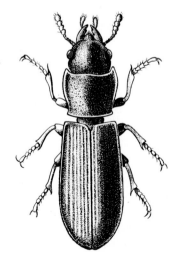

Ostomatidae. Cadelle, Bread beetle in U.S.A., Bolting cloth beetle. A widely distributed pest of warehouses, mills (where it damages the bolting cloth of machines), bakeries and ships, feeding on flour, grain and occasionally on packaged spices, nuts and dried fruits. Also predacious on other insect larvae.

A large, flattened, black beetle (10 mm long) with a distinct waist between thorax and abdomen. Larva (15 mm long) dirty white with black head and prothorax, and also two black 'horns' on end of abdomen. Pupation often occurs in woodwork. The minimum period from egg to adult is 2–3 months but under normal warehousing conditions it may be nearer 9 months. Adults are long-lived (1–2 years) and can survive long periods without food.

Tergite. (*ent*) A cuticular plate forming the dorsal surface of each segment of an insect.

Termite control. (*proc*) Methods for the eradication of termites in infested buildings. Also the protection of building timbers against subsequent attack. The principal commercial methods are:

1) *Site pre-treatment against subterranean termites* (soil treatment, soil poisoning): the application of an insecticide (such as aldrin, dieldrin, chlordane or heptachlor), usually an emulsion in water, to the area on which a building is to stand (Fig. 88). This involves the treatment of all excavations, footings, soil infills and hard-core. The purpose is to provide an insecticidal barrier between the soil and the timber used in construction.

2) *In-situ treatment against subterranean termites* (soil treatment):
   a) *building with underfloor crawl space*; the application of an insecticide (as above) in water emulsion to the whole soil area beneath the building, with special attention to the bases of walls and any bridging structures.
   b) *building with slab floor*; the injection of insecticide through drilled holes in the slab to obtain as even a distribution as possible in the soil beneath (Fig. 87). Also the drenching of the soil in trenches around the outer walls and treatment of the infill.
   In both cases, the horizontal drilling and injection of walls may be necessary, depending on their structure.

3) *Fumigation against drywood termites:* the enclosure of a building in gas-proof sheets and treatment with SULPHURYL FLUORIDE (*q.v.*) or METHYL BROMIDE (*q.v.*): this gives no protection against subsequent termite attack (Fig. 86).

4) *In-situ treatment against drywood termites:* the drilling and injection of timbers with organic solvent preservatives containing an organochlorine insecticide (see above) followed by the brushing or surface spraying of the timbers with the same fluid. Sometimes also the application of emulsion creams to the timber surface (e.g. 'Woodtreat' (*p. prod*)) to provide slow penetration of insecticide.

5) *Use of pre-treated timber.* The construction of new buildings from timber treated by vacuum/pressure techniques, with copper-chrome-arsenate water-borne preservatives. Also the use of pre-treated timber to replace badly infested timbers where these are removed.

Termites. (*ent*) See ISOPTERA.

Territory. (*zoo*) The area occupied by a pest POPULATION (*q.v.*) in its nesting and foraging activities, e.g. the area that a pair of birds may treat as their own and guard against intrusion. Territorial behaviour is well defined among rodents; the area is limited by warmth, harbourage, outdoor cover, available food, and population pressure. Some organisms 'mark' their territorial movements by urination (mice) or by trail PHEROMONES (*q.v.*) as by ants.

Test baiting. (*proc*) Census baiting. Use of unpoisoned food to obtain information on the extent and size of a rodent population; usually with a view to planning a programme of control, or for providing information on the results of a baiting treatment. Test baiting is often carried out as a preliminary to the treatment of a sewer system for brown rat control; not less than 10% of sewer entrances should be lifted, to provide information of value. See also LIVER BAITING.

Test methods. (*proc*) Procedures for evaluating the performance of chemical compounds for pest control use. For example:

Oral dosing: administration by stomach tube of known concentrations of a pesticidal compound in a suitable medium to laboratory animals, (usually rats, Fig. 56), to provide information on symptom expression, speed of action and toxicity; a rapid screening technique to give preliminary information on new compounds.

Choice tests: voluntary feeding of laboratory rats or mice on known concentrations of a compound, formulated in a suitable bait base, to provide information on palatability in two-way, or multiple choice experiments.

Topical application: application of test compounds, as droplets of known volume and concentration, in a suitable solvent, on to the cuticle of individual test insects to provide information on the performance of candidate insecticides. Also known as the micro-drop technique (Fig. 94).

Contact tests: the confining of test insects for specified periods on surfaces (glass, hardboard and others), to which known amounts of insecticidal formulations have been applied. This provides information on insecticidal performance: speed of action, flushing action, recovery from sublethal doses, persistence after application, performance on different substrates, surface penetration, and the ease with which the insecticide is picked up by test insects (Fig. 95).

Repellency tests: evaluation of the properties of compounds for their toxicity and repellency when applied, for example, to simulated harbourages in laboratory tests on cockroaches. Also of compounds applied to surfaces in the screening of potential rodent repellents, the TAKE of food on or near the treated surface being the usual measure of performance.

Field trials: evaluation of new or improved formulations of pesticides under natural conditions in infested buildings: for rodenticides, this normally involves pre-treatment and post-treatment baiting, counting or trapping, to provide a measure of population reduction. See also TEST BAITING.

Tetrachlorvinphos. (*chem*) An organophosphorus compound introduced in 1966 by Shell Chemical Co. under the trade names, 'Gardona', 'Rabon' and 'Ravap'. A white crystalline solid, with a very low toxicity to man, and selective activity against moths and flies. Likely to be of increasing value for fly control in dairies and barns, for the control of certain stored food pests, textile insects, ticks, mites and other animal parasites. Of poor activity to cockroaches.

The acute oral LD_{50} (rat) is 4500 mg/kg. Dermal toxicity is also low. Formulations available include emulsion concentrate and wettable powder.

Tetramethrin. (*chem*) A synthetic pyrethroid introduced in 1965 by Sumitomo Chemical Co. under the trade name 'Neopynamin' with marked knockdown action against flies and mosquitoes, good contact action and one of the more promising replacements for pyrethrins. Action is enhanced by SYNERGISM with piperonyl butoxide.

The technical product is a pale yellow crystalline powder with a pyrethrum-like odour; stable, soluble in most organic solvents. The acute oral LD_{50} (rat) is greater than 20,000 mg/kg; non-irritating to the eyes or skin at normally used concentrations. Formulations currently replacing pyrethrins; notably in aerosol packs and in combination with other insecticides.

Textile pests. (*ent*) Insects which, in their larval stages, attack fabrics made of animal fibres (e.g. clothes, blankets and furnishings). Insect pests of textiles are unusual in their ability to digest the protein, keratin, which is the principal constituent of wool. Soiled fabrics are especially liable to attack. Other materials may also be damaged, but not eaten (e.g. cotton and man-made fibres).

The most important textile pests belong to two Orders, the Lepidoptera (moths) and Coleoptera (beetles). The moths are: the Common clothes moth (*Tineola bisselliella*) now of minimal importance in Britain, the Case-bearing clothes moth (*Tinea pellionella*), the Brown house moth (*Hofmannophila pseudospretella*) widely distributed in Britain, and the White-shouldered house moth (*Endrosis sarcitrella*). The beetles are the Varied carpet beetle (*Anthrenus verbasci*) and the Fur beetle (*Attagenus pellio*), both increasing in importance in Britain in recent years. Occasional damage to textiles may also result from attack by rats and mice.

Thallium sulphate. (*chem*) An inorganic compound used principally as an acute rodenticide; also for mole control, and once finding application in poison baits against indoor ants. A white crystalline solid, readily soluble in water. Tasteless and odourless: a slow acting cumulative poison readily absorbed through the skin. Restrictions apply to its use in many countries (banned in the U.S.A.).

The acute oral LD_{50} (rat) is 16 mg/kg; it is potentially one of the most dangerous chemicals used in pest control because of its high toxicity and cumulative action. The fatal dose for man is less than 500 mg. Early symptoms of poisoning include loss of hair.

Formulations include liquid baits (1%) and the solution absorbed into bread and cereals (0·5%).

Thanite. (*p. prod*) See ISOBORNYL THIOCYANOACETATE.

Thaumatomyia notata. (*ent*) Diptera: Chloropidae. Yellow swarming fly. Small (3 mm long) yellow flies with black bands along the thorax and across the abdomen. They seek the shelter of houses in autumn for hibernation, often swarming close to buildings, entering through windows and causing annoyance in rooms. Two generations a year in Britain. See CLUSTER FLIES.

Thermal fogging. (*proc*) Production of an aerosol by introducing an oil-based insecticide into a heated chamber so that the high temperature causes the insecticide to vaporise and immediately recondense as a fine fog. See FOGGING MACHINE.

Thermal insulation. (*bldg*) The use of materials to conserve heat, e.g. the lagging of water pipes and internal linings to wall surfaces of buildings, cold stores and refrigerated vehicles. These practices increase the possibility of infestation; lagging of pipes in roof voids provides food for textile pests (see HOFMANNOPHILA and ANTHRENUS) and their easy access to the airing cupboards of domestic properties. Insulation boards (especially polystyrene) provide nesting material for mice; the cavity formed by internal linings creates rodent and cockroach harbourage difficult to inspect and treat.

Thermal vaporising unit. (*equip*) T.V.U. An automatic insecticide dispenser: an electrically operated wall-mounted unit consisting of a small heating element and a container for vaporising a solid or liquid insecticide (sometimes fungicide or bactericide) at a regulated temperature and at a constant rate; usually pellets of gamma-BHC, or mixture of DDT and gamma-BHC, sometimes a liquid insecticide impregnated into a solid.

Once used in shops, hotels, canteens and kitchens; now restricted in use because of possible inhalation toxicity to occupants and contamination of exposed foodstuffs. Recommendations for safe use vary with active ingredient. Of value in protecting glasshouse crops. Largely replaced by SLOW RELEASE STRIPS (*q.v.*).

Thermobia domestica. (*ent*) Thysanura: Lepismatidae. The Firebrat (Fig. 90). Usually found near heating plant and around hot water pipes; similar in size and appearance to the silverfish (see LEPISMA SACCHARINA). The firebrat favours higher temperatures than silverfish and damp situations providing high humidities. Numerous moults occur during development (1 year). The adults may live for 2 years and can survive starvation for many months. They are active at night and run swiftly. Primarily nuisance pests but they contaminate their favoured foods (starch, breakfast cereals, paper and glues, and fabrics made from natural and synthetic fibres). Control is achieved using the same insecticidal formulations as for cockroaches.

Thorax. (*ent*) The second, or middle, division of the body of an insect, bearing the legs and (in adults) the wings. Composed of three segments, although frequently only one is visible from above (the PRONOTUM).

Threshold limit. (*tox*) The concentration of a fumigant or solvent vapour in the air above which it is not safe to work. It is usually stated as ppm by volume in air, in relation to maximum exposure periods, repeated exposures during a normal working day and repeated daily exposures. It should be used as a guide for controlling hazards and not as a definite line between a safe and dangerous fumigant concentration.

Thunder-bug. (*ent*) See LIMOTHRIPS CEREALIUM.

Thysanura. (*ent*) Bristle-tails. The Order of most primitive insects containing two pest species, *Lepisma saccharina*, the Silverfish, and *Thermobia domestica*, the Firebrat.

Ticks. (*zoo*) See ACARI, DERMACENTOR VARIABILIS, IXODES RICINUS, RHIPICEPHALUS SANGUINEUS.

TIFA. (*equip*) See FOGGING MACHINES.

Tinea pellionella. (*ent*) Lepidoptera: Tineidae. Case-bearing clothes moth. Not unlike *Tineola bisselliella* in appearance and habits; the wings, however, are darker, the forewings marked by three distinct spots. The larvae construct a case of fibres and silk which is carried by the larva as it moves. The adult does not feed. Development from egg to adult, 10 weeks (22°C).

Tineola bisselliella. (*ent*) Lepidoptera: Tineidae. Common clothes moth, Webbing clothes moth. A pest of fabrics in temperate regions, once more common in Britain than today. Attacks woollens (carpets, upholstery, blankets), yarns, felts, feathers and furs. Larvae prefer textiles soiled with perspiration. Infestations, characterised by silk webbing, occur in the home, in undisturbed cupboards, drawers, under heavy furniture; in manufacturing premises infestations arise in baled goods and wool debris.

The eggs are sticky and adhere to surfaces when laid. The larvae, white with golden-brown head, feed in a silken web, making irregular, large holes in the materials attacked (cf. ANTHRENUS VERBASCI and ATTAGENUS PELLIO). The adult (6–8 mm long) has golden-buff forewings, with no marks; does not feed; the female runs rather than flies; avoids light. Minimum development period from egg to adult is about 6 weeks, usually much longer.

Tobacco moth. (*ent*) See EPHESTIA ELUTELLA.

Tolerance. (*chem*) Maximum permissible level. An accepted residue level of a pesticide in food or animal feed.

Tomorin. (*p. prod*) See COUMACHLOR.

Topical application. (*proc*) See TEST METHODS.

Toxic hazard. (*tox*) The possibility that injury or death to man, domestic animals and other desirable species may result from contact with a pesticide; depending much more on where and the manner in which the pesticide is applied than on how toxic the substance is. Nevertheless hazard usually increases with toxicity of the pesticide: dilution to the concentration for use markedly reduces the risk. Misuse, usually failure to observe elementary precautions given on the label, greatly increases the risk. Hazard to man is greatest in the mixing and application of pesticides, more so than in good secure storage.

Toxicity. (*tox*) The capacity of a pesticide to injure or kill. Highly toxic chemicals, formulated and used at low concentrations, may create a lower TOXIC HAZARD than chemicals of low activity used at high concentrations. Toxicity is measured by the lethal dose (LD) that kills a proportion of test animals (see LD_{50}) according to the portal of entry (mouth, skin or by injection). Rats are usually used for this purpose.

Arbitrary categories of toxicity are recognised:

Category	Acute oral LD_{50} (rat)	Probable lethal dose for man
Extremely toxic	1 mg/kg	a taste, a grain
Highly toxic	1–50 mg/kg	1 teaspoon, 4 cc
Moderately toxic	50–500 mg/kg	1 ounce, 30 g
Slightly toxic	500–5000 mg/kg	1 pint, 250 g
Practically non-toxic	5000–15,000 mg/kg	2 pints, 500 g
Relatively harmless	over 15,000 mg/kg	over 2 pints

Toxicology. (*tox*) The science of poisons. The 'toxicology of a substance' provides information on mode of action, symptoms of poisoning, antidotal procedures and the amounts of that substance which might be hazardous or lethal to man if taken in through the mouth, skin or by inhalation.

Tracheae. (*ent*) Tubes extending from the SPIRACLES (*q.v.*) into the body cavity of insects, carrying oxygen *via* the finer tracheoles directly to the tissues. Also the route of expelled carbon dioxide. Insecticidal fumigants reach the tissues *via* the tracheae and tracheoles by diffusion and active respiration.

Tracking powder. (*chem*) Tracking dust. A powdered substance, usually flour or talc, laid overnight to help determine the presence and location of rodents by foot or tail marks. Valuable in detecting individual survivors following a baiting treatment. The powder is non-toxic and is distinct from a CONTACT DUST (*q.v.*) which incorporates a rodenticide to be picked up on the body and ingested by rodents to kill them.

Transport. (*manuf*) Carriage from one place to another; often of goods in trade, and sometimes of infestation. Use of infested vehicles results in the transfer of pests to the goods being carried and entry into food manufacturers' premises. Especially vulnerable are insulated or refrigerated vehicles which provide harbourage for rodents and insects in the double-skinned walls. Vehicles may be infested prior to loading, having carried another susceptible commodity previously, or having been used previously by industries not so concerned with infestation. CROSS-INFESTATION is not uncommon in the holds of ships and infestation on ships frequently arises from the loading of infested cargoes at ports (Fig. 75).

As part of PREVENTIVE PEST CONTROL (*q.v.*) all transport carrying foods should be scrupulously cleaned, and frequently inspected. Depots used in transport systems are often a source of infestation and these too should receive similar attention. See also SUPPLIER.

Transport of pesticides. (*tox*) Observance in the carriage of pest control chemicals, of regulations concerning flammability and toxic hazard. In practice, the application of common sense precautions to prevent the contamination of goods carried in the same vehicle with pesticides, and to ensure the safety of personnel. In some countries, special regulations may exist.

Vehicles of pest control operators should be so equipped as to prevent the spillage of fluids, and the possible contamination of formulated rodenticides (Fig. 29). Fumigants, fumigation equipment and pesticides with vapour action should be carried in vehicles with separate drivers' cabs.

Trapping. (*proc*) Control of a pest by physical means; the use of a device for killing, or catching alive. Employed primarily in rodent control for removing the last few survivors of a baiting treatment (see also STICKY BOARD) and for bird control, where legislation prohibits direct killing.

Many types of rodent trap are available; 'break back' traps of the treadle type are placed in runs and at the entrances to harbourages. For *Rattus norvegicus*, acceptable baits include meat and fish; for *R. rattus* use fresh fruits, with the traps tied to beams and pipes on well-used runs; for *Mus musculus*, milk chocolate or half a sultana is often successful. Cheese is not particularly attractive to rodents. Baiting of traps may not always be necessary. Traps should be examined daily and reset.

Traps for birds (e.g. for pigeon control) consist of specially designed cages of galvanised mesh into which the birds are lured through tapering tunnels (Fig. 96). Mixed grains are used as bait, the traps being placed on flat roofs in cities, and if used in sufficient numbers can markedly reduce a pigeon population over a period of a few weeks.

Tribolium. (*ent*) Coleoptera: Tenebrionidae. The most numerous of stored product insects, widely distributed and of major importance; occurring in a wide range of commodities (cereals, pulses, ground nuts, dried fruits, spices) and situations: larders, store rooms, warehouses, mills, food manufacturing premises, bakeries, shops and the holds of ships. Heavily infested foods acquire a sour pungent odour.

Small (3–4 mm long) red-brown beetles, very active, long lived (1½ years) but requiring heated premises to survive the U.K. winter.

Tribolium confusum: the Confused flour beetle. A pest of mills, warehouses and bakeries, more resistant to cold than the following species. The eggs are covered with a sticky deposit to which flour particles adhere. The larvae, yellow-brown (6 mm long), active. Minimum period from egg to adult is 20 days (35°C and high humidity); five generations a year may occur in heated buildings. The club of the antenna not so pronounced as in *Tribolium castaneum*.

Tribolium castaneum: The Rust-red flour beetle (Fig. 93). The principal insect pest of ships carrying raw or processed foods. Also in flour mills and bakeries. Similar in appearance and habits to *T. confusum*; adults with a distinct three-segmented club to the antenna.

Trichinella spiralis. (*dis*) See TRICHINOSIS.

Trichinosis. (*dis*) A disease caused by the minute round worm, *Trichinella spiralis*, found in the intestine and spread by both rats and mice. Pigs become infected by eating infested rats and the disease is then transferred to man. The female round worm produces living larvae which migrate through the wall of the intestine to the blood stream and then to various muscles. Here they form cysts which are eaten by man in pork. Mortality in cases of trichinosis is high.

Trichlorphon. (*chem*) An organophosphorus compound introduced in 1952 by Bayer under many trade names including 'Dipterex', 'Tugon' and 'Dylox' (for agriculture). A white to pale yellow crystalline solid of low volatility with a slight but pleasant odour. Trichlorphon has an unusually high solubility in water (15%) but slowly decomposes; highly unstable in contact with alkali, e.g. whitewashed and lime-washed walls and concrete.

Trichlorphon is most effective as a stomach poison, available as a 1% sugar bait (for fly, cockroach and cricket control), which is sprinkled in infested areas, producing rapid knockdown. It has a short life (1 week) when used as a spray.

Of low mammalian toxicity; acute oral LD_{50} (rat) 500 mg/kg with extremely low dermal hazard. POTENTIATED (*q.v.*) by admixture with malathion. Fly baits used outdoors are of low hazard to bees. Formulations available include an 80% soluble powder for use as a water spray and 1% sugar baits.

Trigonogenius globulus. (*ent*) Coleoptera: Ptinidae. Globular spider beetle. Not a serious pest of stored products but occasionally found in warehouses, scavenging on residues. Frequently associated with textile manufacturing, rarely in homes. Typical spider beetle appearance; characterised by light brown elytra with darker spots.

Triton. (*chem*) See EMULSIFIER and WETTING AGENT.

Trogoderma. (*ent*) Coleoptera: Dermestidae. A genus of beetles containing *Trogoderma granarium*, the Khapra beetle (Fig. 100), a most serious pest of grain, adapted to live in hot dry conditions (e.g. maltings). The beetle (2–3 mm long) is ovoid, dark brown and covered with a mottling of lighter brown hairs on the elytra; does not fly and lives only 10 days (35°C). Eggs are laid singly on grain. The yellow-brown larva (7–8 mm long) has bands of darker hairs across the abdomen. Larvae may remain in DIAPAUSE (*q.v.*) for many years. Minimum period of development from egg to adult is 3–4 weeks (35°C); breeding occurs only above 20°C. Control is by fumigation; infestations can persist in small numbers for many years.

Other species of *Trogoderma* include, *T. parabile* a pest of stored grain and cereal products, *T. ornatum* the Cabinet beetle, *T. inclusum*, the Larger cabinet beetle, and *T. versicolor*, a pest of furs.

Trolene. (*p. prod*) See FENCHLORPHOS.

Tropical rat flea. (*ent*) See XENOPSYLLA CHEOPIS.

Fig. 100 Larvae of *Trogoderma granarium* on barley.

Tropical warehouse moth. (*ent*) See EPHESTIA CAUTELLA.

Tropital. (*chem*) A recently discovered synergist for pyrethrum. Tropital is incorporated at ratios between 1:4 and 1:8 of insecticide to synergist. Equal in effectiveness to PIPERONYL BUTOXIDE (*q.v.*) against *Musca domestica*, *Periplaneta americana* and insects of stored foodstuffs; not as effective against *Blatta orientalis*, but superior for *Blatella germanica*. Unstable in sunlight and in the presence of acids.

Tugon. (*p. prod*) See TRICHLORPHON.

Tyroglyphus farinae. (*zoo*) See ACARUS SIRO.

Tyrophagus. (*zoo*) Acari: Tyroglyphidae. A genus of mites (0·3–0·7 mm long) including the Cheese mite, *T. casei*; also *T. longior* and *T. putrescentiae*. Occur on various foods, grains, flour, meats, especially under damp conditions conducive to mould growth. May cause skin reaction in sensitive people. Development is rapid: about 2 weeks (25°C).

U

Ultrasonics. (*phy*) See HIGH FREQUENCY SOUND.

Ultra violet light. (*equip*) Part of the electromagnetic spectrum, in the range 100–400 nanometers, just beyond the blue end of the visible light spectrum and attractive to many pest insects. Exploited in equipment for fly control (Fig. 67); the wavelengths used (300–400 nanometres) are outside the range for bacterial disinfection and do not cause skin or eye irritation. See also PHYSICAL CONTROL METHODS.

U.L.V. (*proc*) Ultra low volume. A spraying technique, initially developed for agriculture, whereby an insecticidal fluid of relatively high concentration is dispersed in oil or other carrier as very small droplets (1–15 microns). The volume of spray applied is very low (about 0·5 g/m³) and forms an ultra fine mist. Suitable for space treatment, reducing application time, degree of wetting and carrier cost. Not recommended for the treatment of harbourages unless the spray is directed into them. Full-face gas masks must be worn.

Unfit for human consumption. (*name*) Condemned as unsuitable for human food because of spoilage organisms, infestation or contamination. Unsound food dealt with by officers of a local authority must be voluntarily surrendered (or otherwise seized): it then becomes the responsibility of the authority to destroy or dispose of the food, so preventing it being offered for sale, or consumed by man.

Urine. (*zoo*) Waste substances of the body dissolved in water, excreted by the kidneys of mammals and birds. The function of the kidneys is to separate these substances from the blood stream, their actual formation taking place in other parts of the body (e.g. liver, muscles). The kidneys also remove excess water.

Most pesticides taken into the body are eliminated by way of the urine. The urine of rats provides a medium for the transmission of *Leptospira* (see LEPTOSPIROSIS). The fouling of food by rodent urine is contrary to Public Health Acts.

Valone. (*p. prod*) PMP. An anticoagulant rodenticide of the indane-dione type; of low palatability to *Rattus norvegicus* compared with warfarin, and less effective as a rodenticide.

Vapona. (*p. prod*) See DICHLORVOS and SLOW RELEASE STRIP.

Vaporiser. (*equip*) See FOGGING MACHINES, SLOW RELEASE STRIP, SMOKE GENERATOR and THERMAL VAPORISING UNIT.

Vapour pressure. (*phy*) The pressure exerted by a vapour above a liquid or solid. When a liquid pesticide is enclosed in a container, the pesticide evaporates until the vapour reaches the saturated vapour pressure, i.e. when no more will evolve and the vapour is sufficiently concentrated to exist in equilibrium with the liquid form of the pesticide. If the pesticide is exposed to air, it will try to attain the saturated vapour pressure but because the air is constantly displaced, continued vaporisation occurs. Vapour pressure increases with temperature. Rate of vaporisation increases at high elevations, i.e. at low atmospheric pressures.

Varied carpet beetle. (*ent*) See ANTHRENUS VERBASCI.

Vector. (*dis*) An organism carrying another which is pathogenic to animals, man or plants, either by accident (e.g. cockroaches, houseflies) or as a specific vector (e.g. certain mosquitoes for malaria). The majority of vectors are insects, and include certain ectoparasites, such as fleas (see MURINE TYPHUS, PASTURELLA), and ticks (see RICKETTSIA).

Vespidae. (*ent*) The family of the Hymenoptera containing the social wasps (yellow jackets) and hornets. Variously marked, the abdomen almost always with characteristic banded colouring of yellow, white, orange or red and black (Fig. 98). Colonies living in nests of 'paper' (Fig. 99), founded by a single queen. Foraging workers are often pests of preserve manufacturing premises (e.g. species of *Vespa* and *Vespula*), attracted by the odour of processing, often contaminating products and causing absenteeism by stinging factory staff.

Vibrissae. (*zoo*) Whiskers. Sense organs of rats and mice used to orientate them in their environment; providing a means by which rodents relate their position and movement to the surfaces with which they come into contact.

Vikane. (*p. prod*) See SULPHURYL FLUORIDE.

Vinegar fly. (*ent*) See DROSOPHILA.

Virginia Polytechnic Institute. (*name*) V.P.I., Blacksburg, Virginia, U.S.A., at which much of the studies for the American NATIONAL PEST CONTROL

ASSOCIATION, on the effectiveness of new insecticides for the control of resistant cockroaches has been carried out in recent years.

Virus. (*dis*) An agent not visible with the ordinary light microscope, often pathogenic to man, animals (and plants), differentiated from bacteria by their smaller size and the fact that they cannot multiply outside living cells. Like bacteria, however, viruses give rise to antibodies in animal serum. They include RICKETTSIA (*q.v.*) and bacteriophages. Causative agents of yellow fever (see AËDES), rabies, poliomyelitis and many other human infections.

Viscosity. (*phy*) A measure of the flow characteristics of a liquid; formulations with high viscosity used in pest control are GELS, TACTILE BIRD REPELLENTS, and the 'adhesives' used on STICKY BOARDS (*q.v.*).

Void. (*bldg*) See ACCESS PANEL, CAVITY WALL and ROOF VOID.

Voles. (*zoo*) Mouse-like rodents, living outdoors, causing damage to agricultural and horticultural crops, attacking the roots of plants, sometimes damaging young trees, but rarely eating and contaminating stored foods. The commonly occurring voles in Britain are the Bank vole (*Clethrionomys glareolus*) with burrowing habits, the Field vole (*Microtus agrestis*) which produces runs below and above ground, and the Water vole (*Arvicola terrestris*), the inoffensive 'Water rat'.

Volume/volume. (*chem*) The ratio of the volume occupied by a component in a mixture, or formulation, to the volume of the total. Usually expressed as, e.g. 10% v/v. See also WEIGHT/WEIGHT.

Vomit. (*ent*) Saliva containing regurgitated, partly digested food. The so-called 'vomit marks' often left by cockroaches as brown smears on walls and infested goods are usually regarded as evidence of regurgitation, but observation shows that this liquid is of faecal origin. The same probably applies to the small round marks left by flies.

W

Warehouse moth. (*ent*) See EPHESTIA ELUTELLA.

Warehousing. (*proc*) The provision and management of clean, dry and well-ventilated covered accommodation for the storage of goods; these conditions alone, however, do not guarantee freedom from infestation.

Cramped storage should be avoided; ideally there should always be ample space for the quantity of goods held, with adequate access and light. Allocation of storage should be made in advance of deliveries so that the space can be cleaned and inspected first. The goods themselves should receive detailed inspection and if necessary be fumigated before arrival.

Stacks should not be built against walls or into corners; they should be kept clear of floors, windows and ventilators. This is a prerequisite for any stack intended for fumigation; failure to comply makes it impossible to cover the stack with gas-proof sheets and difficult to seal effectively if the stack is built around a stanchion.

The best practice is to mark out the stacking plan with white painted lines on the floor, leaving 2 feet around each stow for cleaning purposes. The same procedure should be adopted in corners and at wall-floor angles since this encourages effective cleaning in the very places where residues accumulate and where the first signs of infestation are likely to occur. See also PALLET, STOCK ROTATION and SEGREGATION.

Warfarin. (*chem*) The most widely used anticoagulant rodenticide, of the hydroxycoumarin type; developed by the Wisconsin Alumni Research Foundation, internationally approved; came into use in the U.K. in the early 1950's. Early concentrations (usually 0·005% in ready-to-use baits) have been increased to 0·05% with particular improvement against mice, and in some applications to 0·2%. Much of this increase is associated with the growing resistance to warfarin by rodents in the U.K. over the last 5 years. Most effective against *Rattus norvegicus*; less effective against *R. rattus* and *Mus musculus*.

Like most ANTICOAGULANT RODENTICIDES (*q.v.*), repeated feeding of warfarin over about 5 days is required to kill; does not induce bait-shyness. Long regarded as the 'safe' rodenticide because of the need for repeated feeding, but cats and dogs may be killed by warfarin, as they may occasionally consume large amounts in one feed. Pigs are especially susceptible.

The chronic oral LD_{50} (rat) is 1 mg/kg per day (for 5 days); the acute oral LD_{50} (rat) is about 60 mg/kg. Formulations include concentrates for preparing ready-to-use baits, liquid concentrates (for dilution in water), contact dust (1%) and gel.

Warning gas. (*chem*) A gas often mixed in low proportion with other fumigants to alert operators and others in the vicinity of use. The most important warning gas is CHLOROPICRIN (*q.v.*). It should be recognised, however, that warning gases may not have the same properties as the major toxicant in the commodity being fumigated, with the result that a false sense of safety can be created.

Wasps. (*ent*) See VESPIDAE.

Waste disposal. (*proc*) See REFUSE DISPOSAL.

Water dispersible powder. (*chem*) See WETTABLE POWDER.

Webbing clothes moth. (*ent*) See TINEOLA BISSELLIELLA.

Weevils. (*ent*) See CURCULIONIDAE.

Weight/volume. (*chem*) The concentration of a component in a solution, usually expressed as its weight in grams, in 100 ml of solution. Thus 10 g in 100 ml = 10% w/v.

Weight/weight. (*chem*) The ratio of the weight of a component in a mixture or formulation, to the weight of the total. Usually expressed as, e.g. 10% w/w. See also VOLUME/VOLUME.

Weil's disease. (*dis*) See LEPTOSPIROSIS.

Wettable powder. (*chem*) Water dispersible powder. A very fine powder, usually containing 30–50% of active ingredient (most often an insecticide), readily wetted by water to form a suspension requiring agitation at intervals to prevent settling. The preferred formulation of an insecticide for use on porous surfaces such as rough brickwork (Fig. 101), unpainted timber and concrete; the water is readily absorbed leaving all the wettable powder on the surface, usually resulting in a visible white deposit but readily picked up by crawling insects. The most effective and often the cheapest form of spray. Cf. EMULSION CONCENTRATE and OIL SPRAY.

Wetting agent. (*chem*) Surfactant. A substance which promotes the wetting of surfaces, assisting penetration into fibrous materials, such as textiles and carpets when lightly misted. Commonly used in water sprays, in which soft soap once served the purpose. Lissapol and Teepol are now generally used. Proprietary products are marketed under the name 'Triton'.

Whiskers. (*zoo*) See VIBRISSAE.

White ants. (*ent*) See ISOPTERA.

White arsenic. (*chem*) See ARSENIC TRIOXIDE.

White-bellied rat. (*zoo*) See RATTUS RATTUS.

Wood-boring weevils. (*ent*) See EUOPHRYUM CONFINE.

Wood cockroaches. (*ent*) See PARCOBLATTA.

Woodlice. (*zoo*) See ISOPODA.

Wood mouse. (*zoo*) See APODEMUS SYLVATICUS.

Woodworm. (*ent*) A general term for wood-boring insects, most often

Fig. 101   Treatment of cockroach harbourages with wettable powder in a duct carrying central heating pipes.

beetles, and usually their larval stages. In the U.K., an imprecise term for the larvae of ANOBIUM PUNCTATUM (*q.v.*).

Woolly bears. (*ent*) Larvae of ANTHRENUS and ATTAGENUS (*q.v.*).

Worker termite. (*ent*) A caste of termites, usually the most numerous individuals of a colony and the first to be produced after pairing; similar in appearance to the reproductives, but smaller, sterile, wingless and blind. Often unpigmented (Fig. 16). Responsible for nest building, foraging, and the feeding of nymphs, reproductives and soldiers by trophallaxis: mouth to mouth (stomodeal feeding) or anus to mouth (proctodeal feeding).

World Health Organisation. (*name*) W.H.O. A specialised agency of the

United Nations whose activities are concerned with various aspects of world health, including: collection and dissemination of epidemiological information on outbreaks of disease (e.g. plague); participation in technical work of international significance and the undertaking of surveys by medical experts. In pest control, W.H.O. have published specifications (standards) for pesticide formulations, for spraying and dusting equipment, methods for the control of insects in aircraft, and the drawing up of international sanitary regulations. It has also published standard test methods for use in the laboratory and the field, for detecting and measuring resistance to insecticides.

X

Xenopsylla cheopis. (*ent*) Siphonaptera: Pulicidae. The Oriental rat flea, Tropical rat flea. A cosmopolitan species; a parasite of commensal rats and other wild rodents, remaining on the host's body for long periods; readily attacks man, accounting for its efficiency as a vector of plague (see PASTURELLA PESTIS); also a carrier of MURINE TYPHUS (*q.v.*).

Xestobium rufovillosum. (*ent*) Coleoptera: Anobiidae. Death watch beetle. A well-known pest of old properties, the beetle attacking hardwoods, usually oak (*Quercus* spp.), previously infected by a fungus. Of relatively minor importance economically compared with ANOBIUM PUNCTATUM (*q.v.*). Adults produce a tapping sound by knocking the head against timber, bringing the sexes together for mating.

The adult (6–8 mm long) is brown with patches of yellow-grey hairs; capable of flying but rarely does so; lives for 9–10 weeks but does not feed. Eggs are laid in crevices of timber and hatch in 2–8 weeks. Newly emerged larvae crawl over the wood surface before entering the timber (cf. ANOBIUM). They are 11 mm long when fully grown, white with long yellow hairs, and bore in timber for 3–10 years (av. 4–5 years). Pupae are formed in August (U.K.) and the adult emerges in 2–3 weeks, remaining in the timber usually until the following April–June.

Tunnelling by the larvae is similar to *Anobium punctatum*, the bore-dust is coarse consisting of bun-shaped faecal pellets. Exit holes of the adult are about 3 mm diameter.

Xylene. (*chem*) A liquid distillate of coal tar; of similar properties and uses to BENZENE (*q.v.*) but of lower volatility.

Y

Yellow jackets. (*ent*) See VESPIDAE.
Yellow mealworm beetle. (*ent*) See TENEBRIO MOLITOR.
Yellow swarming fly. (*ent*) See THAUMATOMYIA NOTATA.

Z

Zectran. (*p. prod*) An insecticidal carbamate (with no common name) introduced in 1961 by Dow Chemical Co. which has found little application in industrial and domestic pest control except for limited use against snails and slugs (at 0·5 g a.i./litre). Highly toxic, the acute oral LD_{50} (rat) is between 15 and 63 mg/kg. Available formulations include wettable powder and emulsion concentrate.

Zinc phosphide. (*chem*) A long used acute rodenticide; a dense grey powder, stable when dry, but decomposing slowly in moist conditions to release phosphine, with a strong, pungent (garlic-like) odour. Baits remain toxic for long periods especially when formulated in fats and oils. PRE-BAITING (*q.v.*) is necessary for effective control of rats; it is more readily taken by mice without prebaiting.

The acute oral LD_{50} (rat) is 45 mg/kg; quick acting, confined to professional use in some countries. Ready-to-use baits contain about 2% active ingredient.

ABBREVIATIONS OF ORGANISATIONS

A.P.A. The Association of Public Analysts, 16 Southwark Street, London, S.E.1.

A.P.H.I. The Association of Public Health Inspectors, 19 Grosvenor Place, London, S.W.1.

A.R.C. Agricultural Research Council, 160 Great Portland Street, London, W.1.

B.A.M.A. British Aerosol Manufacturers' Association, Alembic House, 93 Albert Embankment, London, S.E.1.

B.C.P.C. British Crop Protection Council, c/o Agricultural Research Council, 160 Great Portland Street, London, W.1.

B.F.M.I.R.A. British Food Manufacturers' Industries Research Association, Randalls Road, Leatherhead, Surrey.

B.I.B.R.A. British Industrial Biological Research Association, Woodmansterne Road, Carshalton, Surrey.

B.P.C.A. British Pest Control Association, Alembic House, 93 Albert Embankment, London, S.E.1. (*q.v.*)

C.I.A. Chemical Industries Association Limited, formerly A.B.C.M., Association of British Chemical Manufacturers, Alembic House, 93 Albert Embankment, London, S.E.1.

E.P.A. Environmental Protection Agency, Office of Pesticides Programs, Washington, D.C., U.S.A.

F.A.O. Food and Agriculture Organisation of the United Nations, Via delle Terme di Caracalla, 00100 Rome, Italy.

F.D.A. Food and Drug Administration, Office of the Commissioner, Washington, D.C., U.S.A.

I.O.S. International Organization for Standardization, 1 Rue de Varende, 1211 Geneva 20, Switzerland.

M.A.F.F. Ministry of Agriculture, Fisheries & Food, Whitehall Place, London, S.W.1.

N.P.C.A. National Pest Control Association, The Buettner Building, 250 West Jersey Street, Elizabeth, N.J. 07207, U.S.A. (*q.v*)

O.E.C.D. Organisation for Economic Co-operation and Development, 2 Rue Andre Pascal, Paris 16e, France.

P.I.C.L. Pest Infestation Control Laboratory, London Road, Slough, Bucks. (*q.v*)

R.S.P.B.   Royal Society for the Protection of Birds, The Lodge, Sandy, Beds.

R.S.P.C.A.   Royal Society for the Prevention of Cruelty to Animals, 105 Jermyn Street, London, S.W.1.

S.C.I.   The Society of Chemical Industry, 14 Belgrave Square, London, S.W.1.

U.N.E.S.C.O.   United Nations Educational, Scientific and Cultural Organisation, 9 Place de Fontenoy, Paris 7eme, France.

U.S.D.A.   United States Department of Agriculture, Pesticides Regulation Branch, Washington, D.C. 20250, U.S.A.

W.H.O.   World Health Organisation, 1211 Geneva 27, Switzerland. (*q.v.*)